How to Think
about Statistics

How to Think about Statistics

Sixth Edition

John L. Phillips

Boise State University

W. H. FREEMAN AND COMPANY

NEW YORK

Text Designer: Diana Blume

Library of Congress Cataloging-in-Publication Data
Phillips, John L., 1923–
    How to think about statistics / John L. Phillips. — 6th ed.
        p.   cm. — (A series of books in psychology)
    Includes bibliographical references and index.
    ISBN 0-7167-3654-3
    I. Title.   II. Series.
    HA29.P517   1999
    519.5—dc21   99-34901
                                                                    CIP

Printed in the United States of America

Fourth printing, 2001

W. H. Freeman and Company
41 Madison Avenue, New York, NY 10010
Houndmills, Basingstoke RG21 6XS, England

In memory of Elaine

Contents

Preface

While I was teaching a class in educational psychology, it occurred to me that it was my students' lack of background in statistics that was blocking their comprehension of some of the concepts I was presenting to them. There was no room in their curriculum for a course in statistics, so I decided to devote two weeks of my course to the concepts that my students needed. I wrote a small book to help them master those concepts.

When *Statistical Thinking* was published, other people began using it, and I discovered that the concepts needed by my students were the same ones that were needed by students in many courses other than mine, and not only in my discipline but in the social/behavioral sciences generally and in various professional curricula—business, education, and social work, for example—that are related to those sciences.

The culture of any industrialized society is suffused with quantitative information. Some quantitative messages are simple and direct; others involve a relatively complicated process of inference. Knowing how to think statistically makes possible the comprehension of both kinds of messages.

College Use

One of two ways in which students can use this new book is as a supplementary text in a course that demands some statistical thinking but does not focus on statistics. The other use is as a self-teaching preparation for a course that does focus on statistics. It has been my observation, and that of my colleagues, that it is possible for a student to complete such a course without every really *thinking* about statistics. Many students learn to do the required calculations but have only the foggiest conception of what the calculations mean. Although it is true that statistical methods courses discuss the logic of statistics, that is not their focus. This book emphasizes the logical structure of statistical thinking and deemphasizes techniques of data manipulation. If you are planning to enroll in a statistics course, you should read this book as soon as possible after enrolling in the course, before you begin reading the course text. Different books use different symbols to refer to the same concepts, but that will present no difficulty once you have mastered the concepts here. (It might confuse you while you are *learning* them.) I recommend that you finish this book before you begin another, but if that is not feasible, do it as soon as possible after enrolling in the course. In either event you will find that course both easier and more rewarding as a result of this relatively small extra effort at the beginning.

Everyday Use

If you are engaged in business or a profession, you probably encounter quantitative information frequently in your work. If you have no training in statistics and lack the time, inclination, or opportunity to take a course, and if your need is for the consumption (as distinguished from production) of statistical information, then *How to Think about Statistics* can provide you with the necessary background.

But it is not only business and professional people who have access to statistical information and the motivation to interpret it intelligently. If you are a consumer in a competitive economy, the interpretations you make of statistical information will affect your economic well-being. As a citizen, your interpretations will help determine your support of candidates and your position on important political and economic issues, regardless of the kind of work you do.

Imagine that one such issue is currently being discussed by your state legislature: To compete with private industry for competent employees, should the state raise the salaries of its employees? The Penny Pinchers faction in the legislature cites an average salary of state employees that is impressively high already, but the Spectacular Spenders faction says the figures are misleading. How could that be? The two factions agree on the correctness of the basic data from which the average was calculated, and an average is an average, isn't it? Not really, and different types of average are appropriate in different kinds of situations; Chapter 3 will show you why.

Or consider this advertisement for a particular make of automobile: "A survey of owner satisfaction reveals that 31 percent more owners of Cadmobiles than of any other car in its class say they expect to choose the same brand the next time they buy a car." The survey actually was done, and it did indeed turn out as reported. Could its implication nevertheless be deceptive? It could unless you can think statistically. After you have read Chapter 8, you will be able to show how this advertiser could deceive persons who don't know how to think about statistics.

In these examples, the data themselves were solid enough; it was the implications drawn from them that were deceptive. This book is mainly about extracting implications from data. There is, however, another source of error in quantitative information. The data themselves can be faulty. There is a virtually unlimited set of possibilities here, but one can serve to illustrate the genre: Conclusions based upon anecdotal data should be approached cautiously no matter what statistical operations have been performed on them. (Chapters 6 and 11 address problems of validity.) Some anecdotal data are less trustworthy than others: The exploits of golfers and fishermen, for example, are suspect when reported by the persons involved, and young children's reports of events in the real world are likely to be liberally laced with fantasy. Statistical thinking cannot protect you from all possible mistaken conclusions; as computer people say, "garbage in, garbage out."

On the other hand, if you want to draw useful implications from data, many pitfalls remain even when the data themselves are sound. It is in surmounting *those* pitfalls that statistical thinking can be exceedingly helpful.

SAMPLE CALCULATIONS

I have stressed the importance of focusing on logic rather than the manipulation of numbers; for most readers manipulations are best reserved for a course in statistical methods. For others, however, it is reassuring to follow each process through numerically. I have, therefore, introduced several sample calculations. Each calculation and a parallel verbal account is sequestered in a box near the corresponding discussion in the text. It will be easy for you either to follow the calculation or not, as you please.

SAMPLE APPLICATIONS

The logic of statistics is fascinating in itself, but most people want to use that logic to solve problems, to follow solutions proposed by others, and if need be to criticize those solutions. With that in mind I have included at the ends of most chapters some opportunities to test those skills. (Suggested solutions are at the back of the book.) As you do so, you may also be testing your depth of comprehension with respect to the logical principles presented in the main text. Failure to apply a principle may indicate shallow comprehension.

For Readers Who Are Familiar with the Fifth Edition

There are changes scattered through this sixth edition. Many are subtle enough to elude casual inspection, but some are more substantive. In Chapter 5 there is an expanded discussion of *standard and derived standard scores;* in Chapter 7 there are new comments on *degrees of freedom* and on *structural similarities between analyses of central tendency and of variability;* in Chapter 9 six new diagrams illustrate the *reliability of an observed difference between two means;* and in Chapter 10 there is further discussion of the *distinction between qualitative and quantitative measurement* and of *testing the null hypothesis in a chi-square problem.* In the Notes in the back of the book there is a brief discourse on the *central limit theorem* and another on *levels of measurement.*

Acknowledgments

The following persons contributed in various ways to this revision: Dr. Jeffrey S. Berman, University of Memphis; Mr. Dan Huff, Boise State University; Dr. John Pfister, Dartmouth College; Dr. Larry Rogien, Boise State University; Dr. Mark Snow, Boise State University; and Dr. Todd Zakrajsek, Southern Oregon University. I am grateful to all of them.

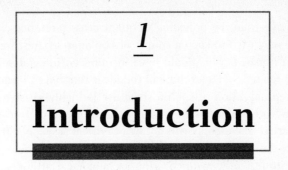

1

Introduction

You may be planning to study statistics not because you want to but because you have to. If so, I know how you feel. I went through the same experience years ago; if I could have avoided statistics, I probably would have. However, my attitude changed after I began to study it, for I discovered in it a new way of thinking that was truly fascinating.

But your present task may be even more challenging than mine was. You won't have to do the computations that I did, but you are about to acquire within a very short time (and possibly by yourself) the same grasp of the underlying structure of statistics that I acquired in two full semesters under an excellent teacher.

The Task

Your plight and your prospects are well illustrated, I think, by the experience of a student who was asked to evaluate the prototype of this book:

> It was the most difficult book I have ever read. It was foreign to me since I had had no background knowledge of the things talked about. . . . If I was to understand it, I realized I would need to outline the book chapter by chapter. I did so, and to my amazement, it followed a very orderly pattern after all. It really did present what the author had stated he hoped to present. If you understood Chapter 1, you could see the logic of Chapter 2, and so on through each chapter. I feel I learned a great deal about measures in a relatively short period of time.

This quotation contains some important advice on how to use the book. I would only add that even though you may have mastered the ideas preceding the one you are working on at any given moment, you should be prepared to go back to those ideas from time to time and consider their relation to the new one being presented. I have tried, by providing frequent cross-references, to help you do just that. (You may wish to keep a couple of bookmarks handy for that purpose.) When you are finished, you should have in mind an integrated structure, with each new idea related to one or more of the ideas that have preceded it.

The conceptualization of such a structure is highly satisfying in itself, but there are many other reasons for making the effort. It is true that an understanding of statistical concepts will not be of critical assistance to you in gathering economic data, in conducting interviews for a political or sociological opinion or attitude study, in uncovering archeological artifacts, or in teaching children. But often people who do economic, political, sociological, anthropological, archeological, or educational studies (to name but a few) report their findings in statistical terms. If you are planning a career in any of the several professions to which those studies are relevant, it is important that you be able to read them with understanding. A continuing awareness of developments in one's field is the mark of the true professional, and this book will help you maintain that awareness.

The Basic Ideas

To understand the meaning of any measurement in the social sciences, you must know at least two things about it. First, you must be able to describe the operations by which it was obtained, and second, you must be able to compare it with other measurements that have been obtained in the same way.

This book is concerned primarily with the second kind of knowledge. Statistical thinking deals with multiple measurements. It analyzes the relation of how many to how much—of frequencies to scores. If the basic element in measurement itself is a *score,* the corresponding concept in statistics is a *distribution* that includes many scores—in short, a *frequency distribution.*

Such a distribution can be described by drawing a picture of it—and indeed in the early stages of your learning about distributions, that is the method I shall use to describe them. The method is cumbersome, however; so ways have been devised to achieve roughly the same result through the use of numbers rather than diagrams. The most important advantage of numbers over diagrams is that they can be manipulated in a way that diagrams cannot.

Imagine that (for reasons known only to your psychoanalyst) you have just had a pile of gravel dumped on your front lawn and that (for similar reasons) you want to describe the result to me over the telephone. To do that successfully, you will have to tell me at least three things about the pile: (1) its general *configuration*—that is, whether it is shaped like a cone, a pancake, or perhaps your garage roof; (2) its *location*—that is, how far and in what direction it is from some reference point familiar to me; and (3) its *dispersion*—the extent to which it is spread out; for example, if it is a cone, is it a steep-sided one that covers only a small area, or is it a low-profile one that covers most of the lawn?

That pile of pebbles is analogous to a *frequency distribution* of scores, and the same kinds of information are needed to describe either adequately. Concerning configuration, we have certain names for frequency distributions that convey information to anyone who is familiar with them—names such as "normal," "symmetrical," "positively skewed," and "bimodal." Concerning location, the procedure is pretty much the same as it is for a pile of gravel: A dimension is identified and a reference point is chosen, and measurement of distance to the center of the distribution is from that reference point. The result is a measure of *central tendency.* Finally, to communicate information about the amount of dispersion, a new reference point is used—namely, the center of the distribution—and the needed information may consist of an average distance of individuals (like that of the individual pebbles in a pile of gravel) from the central point. That distance is a measure of *variability* (dispersion).

But just describing a distribution is not always enough. Often you will be interested in *two* distributions and in the relationship that exists between them. Consider a single variable, "IQ in the general population," and a few other variables with which it might be paired: family income; some index of health care; a general index of socioeconomic status, race, or place of residence. Other interesting relationships could be investigated among the IQs of various subgroups of the general population: between parents and their children, between identical twins, between fraternal twins, between non-twin siblings, and between pairs of unrelated children. These are just a few of the relationships that come to mind at the moment. Others would occur to you if you were making a study of intelligence and its correlates, and in every case you would need a way of communicating your findings to others; in short, you would need a measure of *correlation.*

Whether or not you wish to relate one set of scores to another, you will surely want to be able to tell what each individual measure means. If you are a teacher who has given my child a test, and I ask you how well he did on it, you might try to put me off by saying his score was "high" or "low" and go on to talk about something else. But if I want to know *how* high, you are in trouble. You may answer that he got 90 percent of the test items right. You think you are off the hook, but I persist: "How *hard* is that test? Ninety percent is very impressive if the items are all difficult, but not if most of the other kids score above 95!"

Our conversation has been concerned mostly with the *interpretation of individual measures,* and I'm sure you'll agree that all my questions are pertinent. Without answers to them and others like them, one really cannot know the meaning of a score.

On the other hand, you must be careful to avoid overinterpreting measures, whether they be of individuals or of groups. If, for example, you were to weigh a *random sample**,† of fifty 10-year-old boys, could you easily compute a measure of central tendency for the sample that is identical to the central tendency of a population that includes *all* 10-year-old boys? How different might the obtained weight be if it were computed from a different random sample of that same population? Questions like these have to do with *precision of inference,* which is a kind of *reliability,* and we do have ways of dealing with them.

Assuming for the moment that you do know how to deal with questions about precision, consider this one additional question: Upon further analysis of your sample of 10-year-old boys, you discover a marked difference between the weights of boys living in one area of the country and those living in another. You suspect that the difference is due to diet, and you subsequently narrow that hypothesis down to a single vitamin that seems to be more plentiful in one of the areas than in the other. One way to test your hypothesis would be to select two samples of male infants who are living in the area in which the vitamin is less plentiful, introduce the suspected vitamin into the diet of one group, and then after a period of nine years weigh the subjects in both samples again. Let's imagine that there is a difference between the two. Is it large enough that you can be reasonably sure that it did not occur by chance—that a replication of the same study would not turn up a difference of zero, or even a difference in the other direction? To put it another way, how *significant* is the difference you obtained? Again, there are ways of dealing with such questions.

In the chapters that follow, each of the above ideas will be developed further, but always in the manner that you have seen here. The discussion is aimed directly at the underlying *logic* of statistical thinking, with an absolute minimum of arithmetical and algebraical manipulations. You will find the logic similar in

---

*A random sample is one in which (1) every member of the population has an equal chance of being included in the sample and (2) each selection is made independently of all the others.

†The notes in this book are of two kinds and are denoted by two kinds of superscript. Footnotes are indicated by conventional footnote symbols (*,† etc.); other notes, which appear in the back of the book, are indicated by numbers. The footnotes are intended to supplement the central discussion at any given time and are important to a full understanding of that discussion. A footnote may refine an idea by restricting or extending its implication; it may explain a pedagogical technique by commenting on the relative importance of its various attributes; or it may supply cross-references that are not essential but are enriching. In fact, the objective in every case is enrichment. The way to use the footnotes, then, is (1) to ignore them as you work through a section for the first time and (2) to study them carefully the second time through. Every chapter has a few footnotes. The notes in the back of the book are more technical and go beyond the scope of the text. Ignore them at least until you have mastered the chapter to which they refer.

many ways to common sense. The main difference is that the logic presented here is rigorously systematic, and like any system, its parts are interdependent. So the book cannot be studied piecemeal; the ordering of the chapters is deliberate and necessary. Once you have worked through it, however, the book will serve as a convenient reference for you in your professional life. It has been organized with that in mind.

Description of Data versus Inference to Population

A reader trained in statistics would have noticed in the preceding section a subtle change between the eighth and ninth paragraphs. Every comment before that point was concerned with the *description* of a set of data—its configuration, average value, dispersion, and relation to other data sets. But when we began to consider taking a sample from a population and estimating properties of that population from what we know about the sample, we were shifting from description to *inference*.

That is an important shift—so important that I have devoted a special chapter to it. Chapter 7 is entitled "Description to Inference: A Transition."

Facing Mathphobia

Many persons who are intelligent and who perform many other tasks well find themselves frozen in fear when confronted by any mathematical problem beyond the level of basic arithmetic. If you are not afflicted with mathphobia, skip this entire section and proceed to Chapter 2. If you do have such a phobia, I am not going to attempt to rid you of it; no book is likely to accomplish that. What I believe I can do in this section is show you that your phobia need not be aroused by the contents of this volume, because there is very little mathematics here beyond basic arithmetic—that is, the four fundamental processes of addition, subtraction, multiplication, and division. By "very little" I mean (1) formulas, (2) some arithmetic that is more advanced than the four fundamental processes, and (3) graphic representations of data.

Because most of the concepts in these three categories are or have at some time in the past been familiar to you, the remainder of this section is mostly a very brief review. The one concept that may be new to you (the frequency distribution) is probably the easiest of them all; nevertheless, because it is new, it is treated separately in Chapter 2.

Formulas

You may have thought of formulas as guides to computations: You plug numbers into a formula, follow the rules of algebra, and out comes an answer. Formulas can indeed be very useful in that way, but the emphasis in this book is on the *concepts*—the logical structures—that underlie the computations.

For that reason, the only formulas in the book are definitional. (A *definitional formula* is an equation that defines a concept mathematically.) There are no calculating formulas even where there are calculations, because the computations are there only to illustrate the concepts, and the concepts are illumined by the definitional formulas.

So when you see a new formula, try to discover what it means; look for the relations that it specifies. For our purposes, that specification usually need not be very precise. You may think, for example, of one term in an equation as being "larger" or "smaller" than another instead of "3.14 times as large" or "$\frac{1}{3.14}$ as large." Or you may note that as one term becomes larger, another becomes smaller. For our purposes, that frequently is the most important observation you can make. For example, take the equation

$$v = \frac{d}{t}$$

where v is velocity, d is distance, and t is time. It says that you can find v by dividing the numerator (d) of the expression d/t by the denominator (t). A body that moves a long distance in a given amount of time is moving faster than one that moves a short distance in the same amount of time. Conversely, one that takes a long time to cover a specified distance is moving *less* rapidly than one that takes a short time to cover the same distance. The two sides of any equation are equal by definition; so if there is a change in one side, the other must change, too, in a way that restores equilibrium. Thus, increasing a numerator in a right-hand term has the effect of *increasing* the left-hand term, while increasing the denominator makes the left term *smaller*. To understand basic relationships, this kind of knowledge is really all you need.

If, after you have analyzed a formula in this way, you want to pursue its implications for manipulating data, refer to the "calculation box" that you will find near the text where the formula is introduced. If you want to investigate still further, consult a text on statistical methods. But calculation is not our main concern; for us, it is but another way of illustrating a concept. Our major concern is the comprehension of the concept rather than the calculation of an answer that is numerically correct.

Arithmetic

Concerning the issue of more advanced arithmetic, three concepts will suffice for comprehending the ideas presented in this book: (1) the square, (2) the square root, and (3) negative numbers. If you do suffer from mathphobia but do not have any difficulty with these three concepts, skip the rest of this subsection.

The square of a given number is that number multiplied by itself. If you square 3, you multiply 3 by itself; the result is 9.*

$$3^2 = 3 \text{ times } 3 \quad = \quad 3(3) = \quad 9$$
$$10^2 = 10 \text{ times } 10 \quad = \quad 10(10) = \quad 100$$
$$100^2 = 100 \text{ times } 100 = 100(100) = 10,000$$

The *square root* of a given number is the number that when multiplied by itself produces the given number. After you have squared a number (e.g., $3^2 = 9$), taking a square root of the result ($\sqrt{9} = 3$) gets you back to where you started, namely, 3.

$$3^2 = 3(3) \quad = 9 \quad \text{so} \quad \sqrt{9} = \quad 3$$
$$10^2 = 10(10) \quad = 100 \quad \text{so} \quad \sqrt{100} = \quad 10$$
$$100^2 = 100(100) = 10,000, \text{ so } \sqrt{10,000} = 100$$

Conversely, after you have taken the square root of a number (e.g., $\sqrt{9} = 3$), squaring the result gets you back to where you started (9).

$$\sqrt{9} = 3, \quad \text{so} \quad 3^2 = \quad 3(3) = \quad 9$$
$$\sqrt{100} = 10, \quad \text{so} \quad 10^2 = \quad 10(10) = \quad 100$$
$$\sqrt{10,000} = 100, \text{ so } 100^2 = 100(100) = 10,000$$

In general, since the operations of squaring and taking the square root are precisely the inverse of each other, squaring the square root of a quantity yields the quantity that you started with:

$$(\sqrt{x})^2 = (\sqrt{x})(\sqrt{x}) = x$$

---

*Probably the most common symbol for multiplication is an x interposed between the values that are to be multiplied. When x is also being used to represent a *value*, however, using it to signify an operation (multiplication) can be confusing. One way to avoid the confusion is to let *parentheses* signify multiplication—that is, any value bounded by parentheses is to be multiplied by whatever value is represented immediately adjacent to either parenthesis. Thus $2(2) = 2 \times 2 = 4$, and $4(3 + 17 + 5) = 4(25) = 4 \times 25 = 100$. Also $(2)2$ means the same as $2(2)$, and $(3 + 17 + 5)4$ is the equivalent of $4(3 + 17 + 5)$.

Similarly, taking the square root of the square of a quantity also yields the quantity that you started with[*]:

$$\sqrt{x^2} = \sqrt{xx} = x$$

To put it another way, *the square of $\sqrt{x}$ is x, and the square root of x^2 is also x.*

A *negative number* is opposite in sign to a positive number. When you add such a number to a positive number of the same magnitude, the result is 0. If in Figure 1-1 you think of the horizontal line (the X-axis) as a balance beam with its fulcrum at 0, you can see that a -2 (i.e., negative 2) will balance a $+2$ (i.e., positive 2), a -3 will balance a $+3$, and so on. If you turn the page clockwise 90 degrees, the same will be true of the other axis (the Y-axis).

If the data to be represented do not include any negative numbers—a frequent occurrence—a graph will consist of only the upper right-hand quadrant of Figure 1-1. (That's the part above the horizontal line and to the right of the vertical line.) But if there *are* negative numbers in the data, a surface similar to Figure 1-1 will be necessary to represent all the data graphically.

Graphs

The remainder of this chapter is concerned entirely with graphs. Figure 1-1 lays the foundation for it. Read the following descriptions rather quickly now; then return to them after you have finished the chapter.

You will find two kinds of graphs in this book:

1. When the relation between two quantities (variables)[†] is plotted, the horizontal axis (X) represents changes in one of these quantities, and the vertical axis (Y) represents changes in the other. This is the classical meaning of the term *graph*.

2. When many objects are measured on one dimension X, the numbers of objects at various values of X can be represented as a stack of units at each of the many points of the X-axis. In this usage, Y is the height of the stack at each of those points. This is called a *frequency distribution* (see Chapter 2).

Some mathematicians prefer to reserve the term *graph* for the first of these two kinds of representation, but we shall accept a broader (and more common) meaning and refer to both kinds as graphs.

---

[*]When the terms that constitute an expression are letters rather than numbers, the multiplying operation is implied by the mere juxtaposition of terms: xx means "x multiplied by x," xy means "x multiplied by y," ab means "a multiplied by b," and so on.

[†]A *variable* is a quantity that under varying conditions may have varying values.

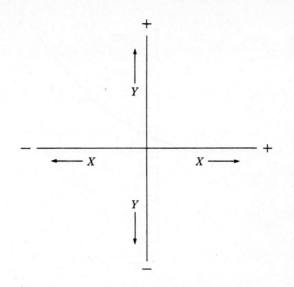

Figure 1-1 Two measures can be represented on a two-dimensional surface by a single point (often called a *data point*). If both of the measures are zero, that point will be precisely where the axis lines cross at the center of the graph.

An example of the first kind of graph is one that represents the relationship between X, the temperature of the air in a room, and Y, the amount of moisture the air will hold—that is, how much water can be added before some of it condenses out of the air. The curve looks something like the one in Figure 1-2. Figure 1-3 shows how performance on a task varies with the performer's level of arousal; there are no negative measures on either axis. In each case, a dot represents a data point, and the simple curve that has been drawn is the one that best fits the points.

The second of the two kinds of graph described above is the frequency distribution. It is the focus of our next chapter.

But before we leave the first kind of graph, I want to warn you against drawing precipitous conclusions from your reading of *any* kind of graph. The arrows in Figures 1-2 and 1-3 indicate the directions in which two variables, X and Y, *increase*. It is possible, however, for the larger magnitudes of a variable to be represented by the smaller number on the graph. If, for example, the Y variable is *skill at golf*, the lower scores (strokes per 18 holes) denote greater skill than do the higher ones. In a case like that, increase in skill could be represented by a falling rather than a rising curve. (The "curve" *could* be a straight line.) For example, if the X variable were *amount of practice* (measured in hours), the graph would look something like Figure 1-4. The relation between X and Y is positive, but it appears to be negative because of the way Y is measured.

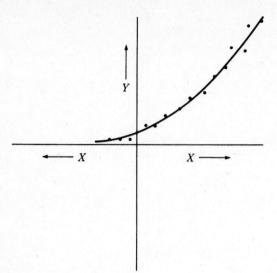

Figure 1-2 Graph of the relation between the temperature of a body of air (X) and the amount of water it will hold (Y). On the X-axis, the zero point is the temperature at which water freezes.

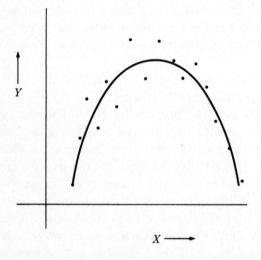

Figure 1-3 Graph showing the relation between level of arousal (X) and performance on a complex cognitive task (Y). There are no negative measures on either axis, and this graph uses only one quadrant (the upper right) of the surface displayed in Figure 1-1.

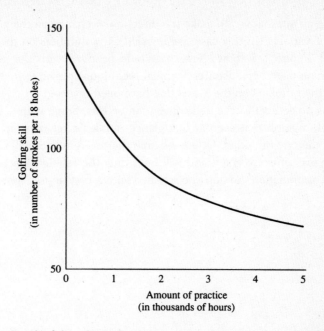

Figure 1-4 Graph of the relation between amount of practice (X) and golfing skill (Y).

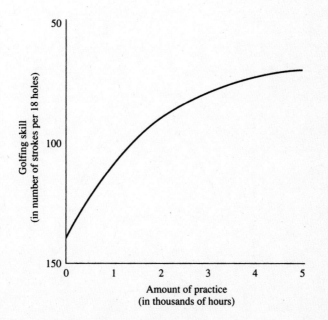

Figure 1-5 Graph of the relation between amount of practice (X) and golfing skill (Y). In order to show golfing skill rising instead of falling with an increasing amount of practice, the golf scores had to be plotted upside down.

There is an alternative. To make the direction of change on the graph reflect that of the *Y* variable (in this case, golfing *skill*), the numbers on the *Y*-axis can be reversed. Figure 1-5 shows how that could be done with the golfing data. Figure 2-11, on page 23, illustrates a similar technique applied to *frequency* data. There, it is the numbers on the *X*-axis that have been reversed.

There is no generally accepted convention on this. Some writers are uncomfortable with a graph that even at first glance implies a relationship that is the opposite of the one intended. Others assume that every reader will carefully examine each axis of every graph and will infer only the relationship that is justified by that examination. So don't be satisfied with a first impression; it could be misleading.

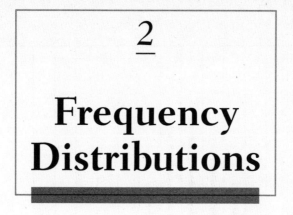

2

Frequency Distributions

Every year, Central University administers a scholastic aptitude test to each of its incoming freshmen. After the tests have all been scored, each freshman is handed a small card with a score on it; then all the freshmen are herded into the football stadium.

A university official stands on the sideline of the field with a microphone in hand. She points to the space between the west goal and the adjacent 5-yard line and announces that anyone with a score below 5 should come down and stand in the middle of that space. Nobody moves, so she walks 5 yards to the east and repeats the instructions for that space. There is a long pause, then one miserable soul slinks down to the field; he is the only one in the *class interval* 5 through 9. Another call (10 through 14) yields 2 students, a fourth (15 through 19) gets 4, a fifth (20 through 24) produces 13, and so on,* with increasing frequencies through the middle scores followed by decreasing ones after that. Figure 2-1 is an aerial view of the field after all the students have assumed their positions. Notice that all the students whose scores are in a given interval (e.g., 15–19) have been placed at the midpoint of that interval (i.e., 4 students at 17). For many purposes we may choose to treat all those scores the same (i.e., as 17s). I shall have more to say about that very soon.

---

*Technically these (5–9, 10–14, 15–19, 20–24, etc.) are all examples of the kind of class interval known as a *score* interval. The section on "Grouped-Frequency Distributions" will show you that the *exact* intervals here are 4.5 to 9.5, 9.5 to 14.5, 14.5 to 19.5, 19.5 to 24.5, and so on up through the highest interval that contains any scores: 64.5 to 69.5.

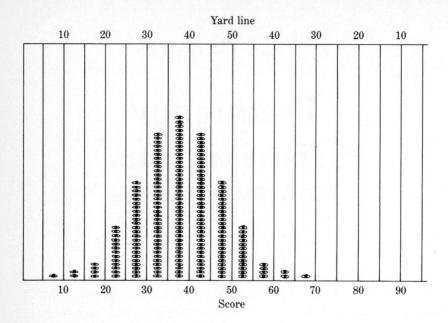

Figure 2-1 Students standing on football field in columns determined by scores on an aptitude test—a frequency distribution.

Normal Distributions

Central U. lacks the budget for it, but if our official had a rope long enough to run from the sideline out around the entire freshman class and back to the sideline, the rope would form what is known as a *normal curve*. A normal curve encloses a *normal distribution*. The "norm" here is a mathematical ideal that is frequently approximated by actual measurements such as the ones taken of CU freshmen. It is *a mathematical model of randomness*. The normal curve is sometimes called a "bell curve" because of the bell-like shape that is clearly discernable in Figure 2-1.

If instead of draping the rope loosely in a smooth curve, our official were to instruct the end student in each column to grasp the rope firmly and pull it taut, the rope would form a series of straight lines and angles known as a *frequency polygon*. Alternatively, we might enclose each column in a rectangle. The resulting figure would be a series of rectangles, each with a width equal to that of the class interval and a height determined by the number of students within that interval—e.g., 2 within the interval 10 through 14, 4 within the interval 15 through 19, and so on. Such a series of rectangles is called a *histogram*.

Whenever measures are arranged in order of magnitude and a frequency is recorded for each magnitude, the result is a *frequency distribution*. The *curve,*

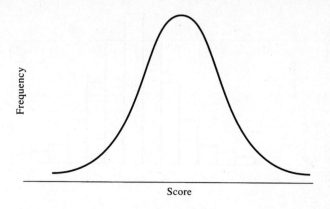

Figure 2-2 Smoothed curve from data of Figure 2-1.

the *frequency polygon,* and the *histogram* are three ways of depicting a frequency distribution graphically (see Figures 2-2 through 2-4).

Quantities that are complexly determined do tend to form normal distributions. Scholastic aptitude measures show that tendency. Such measures have many determiners, both hereditary and environmental; some of those determiners influence an individual's score in an upward direction, some in a downward. In a few individuals, there is a preponderance of positive determiners; theirs are the scores in the upper (right-hand) *tail* of the distribution. In a few others, negative determiners predominate; their scores form the lower tail. Most scores, however, represent more balanced combinations of determiners; they form the large middle area of the distribution.

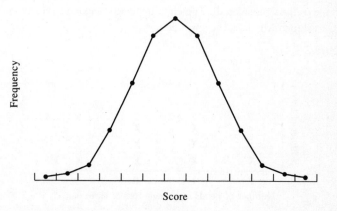

Figure 2-3 Frequency polygon from data of Figure 2-1.

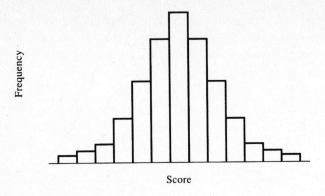

Figure 2-4   Histogram from data of Figure 2-1.

Each combination of determiners is assumed to be a matter of chance, and you can readily see that the probability of, say, 100 determiners being all positive (or all negative) in any individual would be vastly smaller than the probability of there being approximately half and half. If you *don't* readily see that, toss a mere three coins just eight times and plot the number of heads from each toss (0, 1, 2, and 3 are all possibilities).* You should come up with a distribution similar to the one in Figure 2-5 (although with such a small number of observations your distribution might differ markedly from that one). "All heads" (score 3) does not occur often by chance, and neither does "all tails" (score 0). The middle scores are much more frequent.

By the way, do you get the impression when looking at Figure 2-5 that if there were more determiners and more observations, the distribution of coin

*In this example, the three coins correspond to 3 determiners (instead of the 100 cited in the example we just considered). The eight tosses correspond to 8 students instead of the 200 CU freshmen who took the aptitude test.

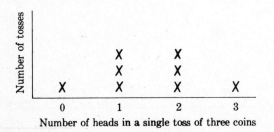

Figure 2-5   Distribution of eight tosses of three coins.

tosses would look very much like the approximately "normal" distribution of students that we got from Central U.? If do you, you are right; the larger those numbers are, the more closely the actual distribution approximates the mathematical model we call the normal curve.

While you have these examples in mind, we can use them to illustrate not only the normality that many distributions share but also something that was first mentioned in Chapter 1: Sometimes we can measure *all* the objects (for us usually people) that we would *like* to measure; sometimes we can *not*. All those objects taken together are called the *population* of interest; if we cannot measure all the objects of our study, that part of the population that we do measure is called a *sample*.

The symbol for the size of a sample is n; for the size of the population it is N. In the case of the CU freshmen, n is 200 if you intend to apply the results of your study to persons outside that group. If you do *not* plan to generalize, that particular class is the *population* of interest, and N is 200. If you *do* generalize beyond that group—say, to college freshmen in general—you will be unable to count or measure the entire population; so anything you say about that population must be inferred from what you know about the *sample*—in this case a sample with an n of 200.

The other example at hand is the coin experiment. There, n is 8, your inference is to all coin tosses, and, again, N is unknown to you, as is everything else about the population. That is, you cannot observe and describe *all* coin tosses; you can only infer properties of such a population from those of the sample that you *do* observe. Our focus in this chapter is on description, not inference; I mention inference here because this section emphasizes normal distributions, and most of the inferential statistics in this book assume the normality of distributions.

Grouped-Frequency Distributions and the Meanings of Scores

One meaning of "score" is a *point* on a scale. But that is an ideal. In practice a measuring operation places an observation within the limits of an *interval*. Probably the most familiar example is chronological age. On your third birthday you were labled a "three-year-old," and that label stayed with you until your fourth birthday. To put it another way, your score on a scale of chronological age stayed with you throughout the interval from your third to your fourth birthday.

Now that system works fairly well, but there would be less error in it if the label "three-year-old" referred not to the interval 3 *to* 4 but to the interval $2\frac{1}{2}$ *to* $3\frac{1}{2}$. Reckoning by the current system, on the day before your fourth birthday the label "three-year old" misrepresented your age by almost a year; by the revised system, no error could be more than *half* a year.

Statisticians use the revised system. Once an observation is known to be somewhere within a specified interval, the observation is treated as if it were at the *midpoint* of that interval even if it is not. On any scale the limits of each interval are halfway from its midpoint to the midpoints of the intervals above and below it: The limit between 5 and 4 is 4.5, and the limit between 5 and 6 is 5.5. So the interval indexed by the number 5 extends from 4.5 to 5.5. If the recorded score is 9, the limits of the interval are 8.5 to 9.5, and so on.

In order to manipulate measures mathematically, we treat them as though they are located at points on a scale; but keep in mind that those points are the midpoints of *intervals*. One reason they should be conceived as intervals rather than points is that many—probably most—of the variables we measure are *continuous* rather than *discrete*. Time passes gradually; the numbers on your digital wristwatch change abruptly. (Time is a continuous variable; the watch's measurement of it is not.) Driving speed changes gradually ("continuously"); "number of speeding tickets" progresses in discrete units. "Scholastic aptitude" varies continuously; scholastic aptitude test scores are discrete, like the numbers on your digital watch.

I shall have more to say later about the continuity and discontinuity of variables and of their measurements. Right now all you need to know is that in many applications, a reported score represents not a *point* on a scale of discrete units but an *interval* on a continuous scale and that it is treated as though it were at the midpoint of that interval.

The conceptually simplest and at the same time most precisely accurate frequency distribution results from merely listing every score that is represented on the baseline and then counting the number of times each of them actually occurs. But doing it that way is like reconnoitering hilly terrain by hiking through its rocks and trees: You get a maximal amount of information, but it is not sufficiently organized to form the kind of *pattern* that you could see easily from an airplane.

Data generated by Central University's freshmen (Figure 2-1) have been organized into class intervals. To see how different the display might have looked if the data had *not* been "grouped," compare Figure 2-1 with Figure 2-6 and Figure 2-7A with Figure 2-7B. Figure 2-6 shows the entire 200 student scores ungrouped. In Figure 2-7A I have magnified a portion of Figure 2-3 (a frequency polygon that corresponds to the distribution in Figure 2-1) in order to emphasize the details exposed by the ungrouped configuration (Figure 2-7B).

If you count the number of students whose scores are in each of the class intervals of Figure 2-7A, you will find just 1 in the class interval 5 through 9. In the 10 through 14 interval there are 2; in the 15 through 19 category, 4; 20 through 24 holds 13 scores; and 25 through 29 has 24. Notice that corresponding intervals of Figure 2-7B contain exactly those same scores. The only difference is that in Figure 2-7A the scores have been "grouped"—i.e., *all* the scores

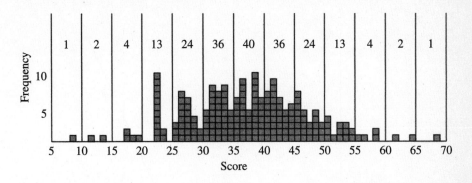

Figure 2-6 Ungrouped scores from which Figure 2-1 was constructed.

in each interval have been moved to the midpoint of that interval, whereas Figure 2-7B gives you the precise location of each score. As you can see, that additional information obscures the general trend of the data.

You may have noticed also that the *numbers* denoting the scores 10, 20, and 30 in Figures 2-7A and 2-7B have been shifted to the right of the *lines* (yard lines in Figure 2-1) that separate one class interval from another. The adjustment is precisely one-half of a unit. That is a refinement of Figure 2-1, because from the beginning I defined the lowest interval as "scores of 5 through 9," the next "10 through 14," and so on. Given that definition, the two limits of each score interval must be (1) below its lowest score and (2) above its highest.

For example, a score of 10 is within the *score* interval "10 through 14," *not* on the line *between* the intervals "5 through 9" and "10 through 14"; so the "10-yard line" in Figure 2-1 lies *below* a score of 10. That line is halfway between the highest score in the lower interval (in this case 9) and the lowest score in the higher interval (in this case 10); the line is therefore actually at 9.5. Accordingly, any score of 10 will be placed *above* the lower boundary of an interval that extends from 9.5 to 14.5, and all other intervals are marked off in the same way (e.g., 4.5 to 9.5 . . . 14.5 to 19.5 . . . 64.5 to 69.5). These are the *exact* class intervals that I promised on the first page of this chapter.

Five illustrations of the distinction between the score limits and the exact limits of a class interval are given in Figure 2-8. Score limits are inscribed above the baseline, exact limits below it.

This diagram also identifies (by means of a small arrow) the midpoint of each interval. Notice that the only midpoints that are whole numbers (41, 42) are in the two intervals (B and C) that are 3 and 5 score units long, respectively; 3 and 5 are odd numbers. Every interval that subtends an *even* number of units (2 in A, 10 in D, 20 in E) has a midpoint that falls *between the middle two* scores (40 and 41, 44 and 45, and 49 and 50, respectively). The midpoints of these

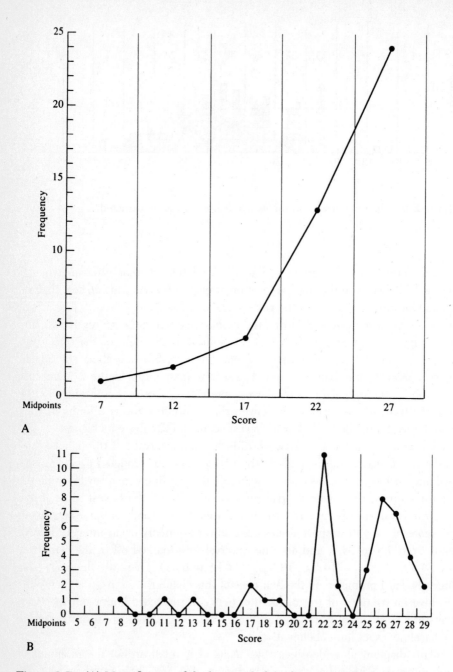

Figure 2-7 (A) Magnification of the lower tail of the frequency polygon depicted in Figure 2-3, which corresponds to the distribution in Figure 2-1. (B) Same set of scores as in Figure 2-7A but without grouping.

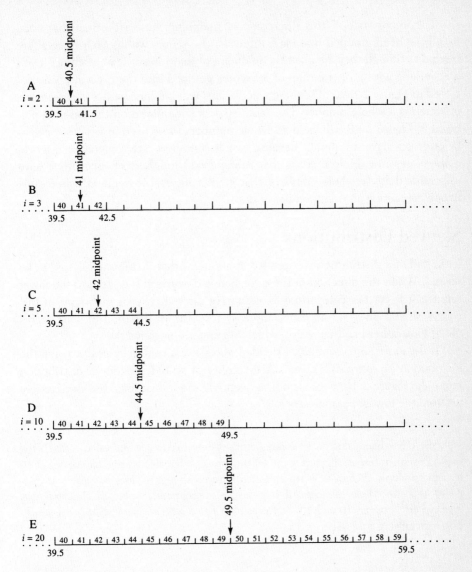

Figure 2-8 Class intervals of 2, 3, 5, 10, and 20. The numbers arranged sequentially from left to right above each baseline are scores. The score limits of intervals, reading top to bottom, are 40–41, 40–42, 40–44, 40–49, and 40–59. Corresponding exact intervals are 39.5–41.5, 39. 5–42.5, 39. 5–44.5, 39. 5–49.5, and 39. 5–59.5. *Midpoints* are 40.5, 41, 42, 4.5, and 49.5.

three pairs of scores are 40.5, 44.5, and 49.5, none of which is a whole number. A close examination of Figure 2-8 should make it clear to you why that is the case. (To make it easy, compare just A and B, which contain only 2 and 3 score units, respectively.)

The importance of this discussion of midpoints lies in the fact that once data have been grouped into class intervals, the scores within each interval are treated as though they are all at the midpoint of that interval. For example, if you are working with a distribution of 30 scores grouped into the class intervals depicted in Figure 2-8A, subsequent calculations will be done with 30 scores none of which is a whole number. The same 30 scores deployed on the baseline pictured in Figure 2-8B will yield 30 whole numbers to be used in whatever calculations you have in mind. Because whole numbers are generally easier to manipulate, and since intervals that subtend odd numbers of score units have midpoints that are whole numbers, that kind of interval is frequently preferable to one containing an even number of units.*

Skewed Distributions

If the tail of a distribution is extended abnormally, that distribution is said to be *skewed*. When the direction of the extension is downward (i.e., toward the lower values), it is further categorized as *negatively* skewed. If your instructor in this course should give you an especially easy test, the distribution of your scores might look rather like Figure 2-9, which illustrates a negative skew.

If the *upper* tail is similarly extended, the skew is said to be positive, and the distribution is *positively* skewed. A difficult test would produce a distribution more like Figure 2-10, because only a talented (and industrious) few deviate very far from the lowest possible score.

---

*On the other hand, our decimal numbering system makes an interval size of 10 the most convenient, especially during the tallying process. For that reason, an interval of 10 or some multiple thereof is worth considering (even though it does not give you midpoints that are whole numbers) if it yields an appropriate number of class intervals (somewhere between 10 and 20) and if you are faced with so many scores that convenience in tallying is an issue.

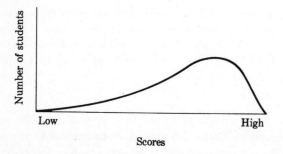

Figure 2-9 Negatively skewed distribution.

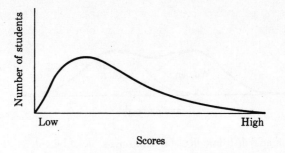

Figure 2-10   Positively skewed distribution.

Other Configurations

There are other possible distributions. If we were measuring conformity behavior, like that of motorists at a busy intersection, we might get a **J** *curve*, like the one in Figure 2-11. The same would be true of a distribution of scores on a test that is extremely easy (so easy that most people get perfect scores) or extremely difficult (so difficult that most of the scores are zero). See also the discussion on "Skewed Distributions."

An entirely different configuration would emerge if we were to measure the standing height of humans, because there are two physical types of humans, male and female. Thus, we should obtain a *bimodal* distribution, as shown in Figure

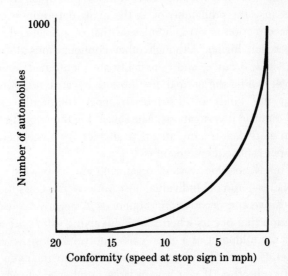

Figure 2-11   **J** curve of conforming behavior.

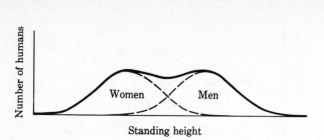

Figure 2-12   Bimodal distribution of humans on a scale of height.

2-12. Similarly, if your instructor were to spring a pop quiz on Chapter 9 of this book at a time when only half of the class had read it, the distribution of those scores would be bimodal. (Note that in Figure 2-12 the distribution is apparently quite symmetrical; however, a bimodal distribution can be asymmetrical, as indeed this one surely would be if, say, it were composed of twice as many women as men.)

The **J** distribution is difficult to deal with statistically, and a bimodal distribution can be dealt with by separating the two normal distributions that are partially concealed within it. So these other configurations are of only passing interest to us here.

SUMMARY

Many frequency distributions in the social sciences are approximately normal; that is, in the typical distribution there are a few very low scores and a few very high ones, but the great mass of individuals tend to pile up on the middle scores. This occurs because the probability of all the many determiners of a trait pointing in the same direction is virtually nil, and that of a balanced combination of determiners is much higher. Although other configurations do occur (positive skew, negative skew, **J** curve, and bimodality are mentioned), our primary concern in this book will be the normal distribution, because it serves as a good approximation to the kinds of distribution most frequently encountered in behavioral and medical investigations. Moreover, it is the configuration for which the best-known statistical treatments are available. We shall encounter some of those treatments in subsequent chapters.

There are many possible ways of organizing and displaying data. One is to "group" frequencies—to pool individual observations into class intervals. Grouping has both positive and negative consequences. A negative consequence is the loss of information that occurs when scores of varying values are assigned whatever value is at the midpoint of a given class interval. A positive consequence is that grouping tends to reveal underlying patterns by allowing many random variations to cancel each other. If you have a large number of observations to organize, another positive effect is that grouping can simplify calculations.

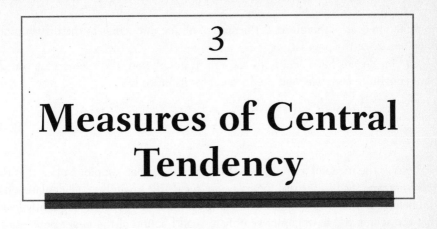

3

Measures of Central Tendency

Somewhere in the intermediate grades you were introduced to the concept of *average,* and you may have used it ever since on the assumption that there is only one such. Actually, there are several types of average, of which three of the most common will be described here: the mean, the median, and the mode.

The Mean (μ and $\overline{X}$)

The average that you learned about in grade school was a *mean.* The mean is not, as you were led to believe, *the* average, but it does have characteristics that make it the best one to use in many circumstances. Whenever the frequency distribution is fairly symmetrical and a calculator is available, the mean is the statistic of choice. Considerable computation time is necessary because all the scores must be added together before the sum can be divided by N (the number of scores), and there are sometimes thousands of scores to add. If you have access to the entire target population, the formula for the mean is

$$\mu = \frac{\Sigma X}{N} \qquad (3\text{-}1)$$

where μ is the mean of the population, Σ is a combining term meaning "sum of," and X refers to the *raw scores* that have been recorded.* The expression ΣX (read

---

*A *raw score* is one that does not imply a comparison with any other score; "inches," "points," "runs," "hits," "errors," and "number right," when those units are simply counted, are raw scores. Raw scores are the only kind we have dealt with so far; others will be introduced later.

"summation ecks"), therefore, is the sum of all the raw scores in the distribution. N is the size of the population.

If you do *not* have access to the entire population, the scores that you *do* have constitute a *sample,* and the formula for *its* mean is

$$\overline{X} = \frac{\Sigma X}{n} \tag{3-2}$$

where $\overline{X}$ is the mean of a sample and n is the size of that sample. Notice that *the operations specified by the two formulas are exactly the same.* If you calculate what you believe to be a population mean using Formula (3-1) and subsequently learn that the scores you have comprise only a small fraction of the target population, all you need to do is to recognize the outcome as a sample mean (and to change two labels: μ to $\overline{X}$ and N to n). No further calculation is necessary.

Because in published research the description of samples is more common than that of populations, every reference to "the mean" after this chapter will be to a sample mean unless otherwise noted (even though your own studies may frequently be of populations).

In any event, Formulas (3-1) and (3-2) say the same thing, and what they say has two aspects. The first is that the formulas *define* the mean; the second is that they specify a procedure for *calculating* it. But this book is not about calculating; our primary objective is rather to acquire an understanding of statistical concepts. The concept of the mean can best be understood through the following illustration.

Essence of the Concept

Try to imagine a long beam made of an exotic metal that is totally rigid and utterly weightless. Upon that beam we shall place 14 cubes (to represent 14 scores), all of equal weight. Figure 3-1 shows one possible distribution of cubes.

At what point on the beam would a fulcrum (support) have to be placed to establish an equilibrium? That is, what is the *balance point* of the distribution? Because this distribution is symmetrical, it should be easy to see that it would balance if the fulcrum were placed at 5.5. But that is what all the computation is

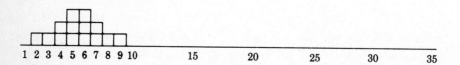

Figure 3-1   Fourteen cubes on a weightless beam: a symmetrical distribution.

about; by using the formula, we can find the precise point we are seeking* even in cases where a diagram would be rather puzzling. (That such situations do arise will become clear in the section on "The Median.")

Another way of saying "balance point" is "the point from which all deviations sum to zero." In Figure 3-1, for example, that point is 5.5. One of the scores (the 2) deviates $3\frac{1}{2}$ units in the negative direction (left) and one (the 9) deviates $3\frac{1}{2}$ units in the positive direction (right); one is a negative and one is a positive $2\frac{1}{2}$; there are two negative and two positive deviations of $1\frac{1}{2}$; finally, there are three scores $\frac{1}{2}$ unit below and three scores $\frac{1}{2}$ unit above the mean. Most distributions will not be perfectly symmetrical, as this one is, but all will produce the same result: The positive and negative deviations from the mean will always cancel each other, bringing their sum to zero.

What I have been saying in different ways is that *the mean is the balance point* of any distribution of scores.

Appropriate Applications

As the balance point of a distribution, the mean is the only measure of central tendency that is sensitive to all its scores. Probably its most important application is to other measures. Because of that balance-point feature, it is compatible with many more complex measures that you will meet later. Indeed, calculating a mean is an integral part of calculating a *standard deviation,* a product-moment *coefficient of correlation,* and all the various *standard errors,* to name a few.

A related advantage of the mean over other measures of central tendency is its usefulness in making inferences from sample to population: The mean of a sample is the best estimate of that of the population. But even the best estimate will probably miss the mark, and it is important to know the probable extent of the error. The mean lends itself to error estimation as well, as you will see in Chapter 8.

So if you want to know the population mean but all you have is a sample, you will probably choose the mean to represent central tendency in your sample. And if you anticipate calculating more complicated statistical measures, you will probably need to calculate your sample mean first.

As I mentioned in the preface, some people do not feel comfortable with a quantitative concept until they have followed the relevant computation. If you are one of those people, Box 3-1 provides you with an opportunity to do that. The calculation is based on the defining equation for the mean. Special calculational formulas frequently differ from definitional ones (see page 6), and when

$$*\overline{X} = \frac{\Sigma X}{N} = \frac{\text{sum of raw scores}}{\text{number of raw scores}} = \frac{77}{14} = 5.5$$

Box 3-1   Calculation of a Mean (see Figure 2-1)

(1) Class interval	(2) Midpoint X_m	(3) f	(4) fX_m
65–69	67	1	67
60–64	62	2	124
55–59	57	4	228
50–54	52	13	676
45–49	47	24	1128
40–44	42	36	1512
35–39	37	40	1480
30–34	32	36	1152
25–29	27	24	648
20–24	22	13	286
15–19	17	4	68
10–14	12	2	24
5–9	7	1	7
		$\Sigma = 200$	$\Sigma = 7400$

Column 1: Class intervals. See Figure 2-1, page 14.

Column 2: Midpoints of class intervals. When data are grouped into class intervals, all individuals (X) within each interval are treated as though they were at the midpoint of that interval (X_m). Of course, that is not strictly true, but if the class intervals are small and the n is large, error is negligible.

Column 3: Frequency (f) of scores in each class interval.

Column 4: Product of midpoint and frequency (fX_m). Only one person scored in the 5–9 interval, and the midpoint of that interval is 7; so the fX_m for that interval is just 7. There are two scores in the 10–14 interval and its midpoint is 12, so its fX_m is $2 \times 12 = 24$. Four persons scored somewhere in the interval 15–19, so fX_m is $17 \times 4 = 68$, and so on through the remaining intervals.

$$\overline{X} = \frac{\Sigma X}{n}$$

It should be clear that the sum of column 3 is n. It should also be clear that simply adding all the individual X_m's will give us the same sum as that of column 4; that is, the sum of column 4 is the ΣX in the formula, give or take an error of negligible magnitude.

$$\overline{X} = \frac{7400}{200} = 37$$

they do they tend to obscure the *meaning* of the operations they represent. Since meaning is your mission here, it will be better for you to keep your calculations close to your concepts.

The Median (Mdn)

Another way of indicating central tendency is to tell which point on the baseline divides the distribution into two equal parts. Note that I said it divides the *distribution* in half, not the *baseline*. Look back at Figure 3-1. There, the beam is the baseline; it is 35 units long, but $\frac{35}{2} = 17.5$ is not the median. Nor is the median defined as a point halfway between the lowest point (1.5) and the highest (9.5), although because of the perfect symmetry of the distribution, it happens to be there in that particular case.

Essence of the Concept

In every case, the median is the point between the lower and upper halves of the distribution. (Remember, the distribution is the group of individual scores comprising the sample.) In Figure 3-1, that point is 5.5, because there are 7 scores below it and 7 above. In any distribution with an N of 14, the median will be midway between the 7th and 8th scores (counting up from the bottom of the distribution); in any distribution of 1000 individuals, the median will be the point midway between the 500th and 501st; and so on, with as many illustrations as you care to cite.[1]

Appropriate Applications

"Who is in what half of the distribution?" If that is our question, the answer will depend upon our first finding the median. A more important characteristic of the median, however, is that although it is not sensitive to the exact location of every score in the distribution, it can be used in situations where the mean would be inappropriate.

It may not be too much of an oversimplification to say that the median should be used in preference to the mean whenever the shape of distribution departs radically from perfect *symmetry*. Consider the situation depicted in Figure 3-1. There, because the distribution is perfectly symmetrical, the mean and the median are at precisely the same place (5.5). Now look at Figure 3-2 to see what happens when that symmetry is disturbed. In that figure we have exactly the same distribution as in Figure 3-1 except that two of the scores have been shifted far to the right. The balance point—that is, the mean—has shifted also; it is now 9.5, which, lying as it does above 12 of the 14 scores, is probably an inappropriate index of the central tendency of this distribution.

But what has happened to the *median* as a result of that shift of two scores? It hasn't moved at all! (Count the scores above and below it, and see for your-

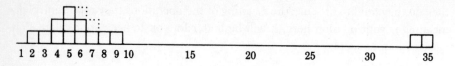

Figure 3-2   Fourteen cubes on a weightless beam: an asymmetrical distribution.

self.) You may say that it *should* have moved—at least a little bit—because the distribution is different from what it was before; however, even though it is insensitive to that change, I'm sure you will agree that in this distribution the median is a better measure of central tendency than the mean, because the two extreme scores have *too much* influence on the mean. Often what has happened in such cases is that all the scores have been forced into one category when they should have been classified into two or more. For example, if you were to plot a distribution of the annual incomes of a football coaching staff, you would probably find that most are rather close together but that the salary of the head coach is distinctly separate from the others. If you had to report a single average of the salaries of football coaches at Central University, would you use the mean or the median? If you *don't* have to confine your report to a single central tendency, it probably would be better to report the head coach separately in this case, but if salaries of the senior members of the support staff approach that of the head coach, the median of the entire staff might be an appropriate index of central tendency.

In another situation that militates against using the mean, the scale is not long enough to accommodate all the scores at one end of the distribution. For example, whereas a long, difficult test might produce a normal distribution of student scores, a short, easy one might pile up, say, a third of the scores at the top of the scale. The true magnitude of these scores is indeterminate: There is no way to know "how far out on the bar" each weight should be placed and hence no way to locate the balance point of the distribution. But you can use the median.

Finally, there is the rather rare array of data that are not precisely quantitative but nevertheless appear in a universally recognized *order*.* Military ranks come to mind, but any ranking structure could illustrate the genre. Consider the results of a race—say, a marathon—in which results are recorded in two forms: (1) the precise *time* that it took each contestant to reach the finish line and (2) the *order* in which they all reached it. If you know the runners' *times*, you can calculate their mean time, but if all you have is their *ranks*, the median will have

---

*For a classification of various kinds of data, see note 1 for Chapter 10, page 175.

to do, because there are no *scores* to support the calculation of a mean. Even a median would not be very edifying, however, unless there really is someone who needs to know the rank of the runner at the middle of the pack.

The Mode

The last of the three averages to be presented here is also the easiest. The *mode* is simply the point with the greatest frequency. In Figure 2-1, the mode is 37; in Figure 3-1, 5.5; in Figure 3-2, 5.0.

If your data are quantitative and ordered, as are nearly all the data presented in this book, the mode's very simplicity explains one of its two main purposes: It is used when a very quick estimate is needed. It is also used specifically for identifying the typical (most common) score.

If your data are qualitative and not ordered, like sales of various colors of designer dresses or enrollments in the subject-matter categories of a college curriculum, the mode is really the only central tendency that *can* be used: You cannot count up from the bottom as you must in order to find a median, much less add scores as you must to find a mean. There *is* no "bottom," and there are no scores to add.

The mode can therefore be used where the mean and the median cannot. Beyond that, it can supplement but not supplant either or both of those statistics.

SUMMARY

The three most common measures of central tendency are the *mean,* the *median,* and the *mode.*

The *mean* is the balance point of a distribution. It is the only average that utilizes all available information, and when distributions are approximately normal it is the one that serves as an essential component of many of the more complex measures that you will encounter in later chapters.

The point at which a distribution can be cut in half is its *median.* It is used when precise location of the two halves is a prime concern, when data are ordered but not precisely quantitative, or when a distribution of precisely quantitative data is far from normal. In those circumstances the median can supplement or even supplant the mean.

The *mode* is the place where the greatest number of cases occur; if the data are quantitative, it is the most common score. When that is exactly the information you want or when your data are neither quantitative nor ordered, the mode is the appropriate index of central tendency. In other situations it can be useful as a supplement to the median and/or the mean, but it should never supplant them.

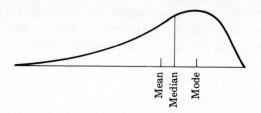

Figure 3-3   Measures of central tendency in a negatively skewed distribution.

In a skewed distribution, the arrangement (order) of the three measures along the baseline is predictable. If the skew is negative, as in Figure 3-3, the arrangement from left to right is: mean, median, and mode. If the skew is positive, as in Figure 3-4, the order is just the opposite: mode, median, and mean. Conversely, if you know the order of the three averages, you can tell the direction of the skew.

Because of the mean's affinity to many of the more complicated constructs that I will present later (beginning in Chapter 4), future references to measures of central tendency will be almost exclusively to the mean. For simply describing a small population, however, or for doing a preliminary exploration of data that might *later* be used to make inferences beyond your sample, all three measures can be helpful, as can graphic representations.

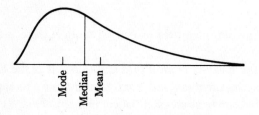

Figure 3-4   Measures of central tendency in a positively skewed distribution.

Sample Applications

The following problems require that you make some decisions about the appropriate use of statistics. In each case, you are presented with a situation that might be faced by a practitioner of the indicated discipline. Imagine yourself as that person in that situation.

Make a tentative choice without looking back into the chapter; explain it as best you can. Then review the text and refine your answer. Finally, check the back of the book (pages 179–195) for a suggested choice and brief discussion of it.

You may use that choice and that discussion to correct or further refine your response. But if you feel that yours is as good as mine, discuss it with your instructor or someone else who has a knowledge of statistics. If he or she agrees, I'd very much like to see what you have done.

Some of the situations described here are adapted from events that actually occurred; others are wholly contrived. In the *Possible Misinterpretation* sections of the "Solutions to Sample Applications" at the back of the book, when I describe an alternative to the earlier interpretation I mean to imply that the alternative is possible—maybe even very plausible—but not that it is necessarily correct.

EDUCATION

A cooperative vocational education program has recently opened to provide one year of training for 200 12th-grade students from 10 schools. The teachers are in the process of selecting and developing curricular materials, but they are unsure of the appropriate reading level for the materials. They decide to get an estimate of the reading skills of the students enrolled in the program, and you are called in as a consultant. You suggest that the teachers administer a standardized reading comprehension test and obtain for each student a single score indicating the reading level of that student. Now what do you do?

POLITICAL SCIENCE

You are studying the domestic and military expenditures of European nations. You want to find the average amount spent by European countries on arms. You have gathered your data. Now what do you do?

PSYCHOLOGY

You are a family counselor working with the mother of a three-day-old infant. The mother is very concerned about her child (her sister has recently given birth to an infant with birth defects) and asks you if her infant is showing normal behaviors for a newborn. You observe the infant in question, but you are not certain which behaviors are classified as normal for a neonate. Your task, therefore, is to find out how newborns tend to behave. You go to three hospitals in your city, visit the neonatal units, and observe and measure a variety of infant behaviors. For example, you dangle a large red ring in front of each infant to see whether it follows the ring visually or even attempts to grasp it. You sound a bell close to each infant's right ear and observe whether the infant turns toward the sound. You then assign points based on your observations (e.g., 1 point for following the ring visually and 2 points for grasping the ring). What single statistic would best represent all the children you have tested?

SOCIAL WORK

The director of a child welfare agency is interested in the length of time that families receive protective services. She asks you to provide information regarding the number of treatment hours received by these clients. You extract the required data from the closed-case files. How can those data be represented statistically?

SOCIOLOGY

A city council wants to know the average income of its residents. Assuming that you have access to the data, how will you analyze it in order to answer the council's question?

4

Measures of Variability

Different populations (and the samples extracted from them) have different central tendencies, but they differ in another significant respect as well. Consider the two curves depicted in Figure 4-1. Both represent distributions of the same area (identical Ns), and both have the same central tendency; nevertheless, the two distributions are very different. In what way are they different? You can see that one is spread out more than the other. Since the baseline on which the spreading occurs is a single scale of scores, the spreading means that the scores in that distribution vary more than those in the "squeezed together" distribution.

Diagrams provide a superior way of approaching a new concept, but even if we had a diagram of every distribution drawn to scale so that we could compare variabilities by inspection, there would still be a need (demonstrated in Chapter 5) for an index that can enter into mathematical operations that are essentially numerical; you can't multiply or divide a visual perception with an acceptable degree of precision. There are many occasions (some of which are discussed in Chapter 5) when it is important to have some kind of numerical index of the variability of a set of scores. In this chapter, we shall examine three such indices.

The Standard Deviation (σ and S)

Taking its importance on faith for the moment, let us consider how an index of variability might be devised if we had none already available. It may help to have a concrete example in mind during this discussion, so let us imagine that instead of the single scholastic aptitude test mentioned in Chapter 2, those students took a *pair* of tests—one of verbal aptitude, the other of numerical. The

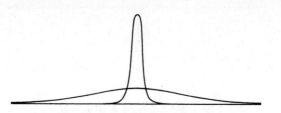

Figure 4-1   Two distributions with the same *N*s but different variabilities.

distributions of scores on the two tests might look something like Figure 4-2.* In that diagram, the means of the two distributions have been aligned so that we may concentrate on their respective variabilities. As we look at the figure, we are immediately impressed by the striking difference between the dispersions of the two distributions across their baselines. But it is not enough to be impressed; we need a numerical index of dispersion (variability).

Essence of the Concept

How might such an index be devised? One way would be to compare every score in the sample to every other and to average the differences obtained thereby. But that would be entirely too cumbersome, especially with large distributions; we can get the same effect by selecting a point of reference in the middle of the distribution and measuring the distance of each individual score from that point, as in Figure 4-3. The average (mean) of those differences (without regard to sign) can also serve as an index of variability. If, for example, we were to choose the mean of our target population as our reference point, our index would be an

---

*The indicated central tendencies and variabilities of the two distributions make somewhat plausible the proposition that if combined, they would form the distribution shown in Figure 2-1. However, that was not the primary consideration in their selection. Rather it was *simplicity.* We are deemphasizing computation; therefore, computations that *are* required have been made as easy as possible. It is easier to think about an interval that extends from 20 to 25, for example, than one that extends from 23 to 27, or even from 23 to 28. You will find that you can do most—possibly all—of this book's computations entirely in your head.

But do not attempt any computations unless the requisite data are readily available. For example, in Figure 4-2 do not attempt to confirm the standard deviations of 15 and 5 in the two distributions. To do so, you would need a list of the individual scores, and those have been withheld deliberately to encourage you to focus on the big picture—the configuration of the entire sample for each test and the comparison of two configurations that are very different from each other.

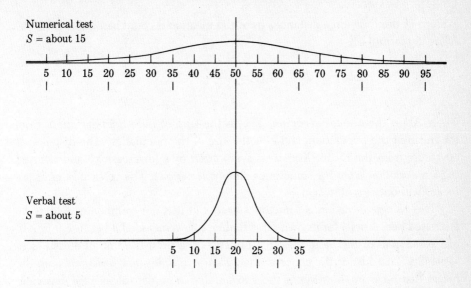

Figure 4-2 Two distributions with the same Ns but different variabilities.

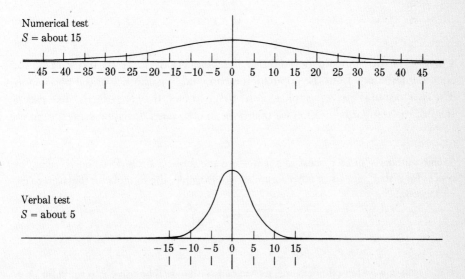

Figure 4-3 The distributions in Figure 4-2 with raw scores converted to deviation scores.

average of the individual distances from the mean and could be obtained via the following formula:

$$AD = \frac{\Sigma |x|}{N} \qquad (4\text{-}1)$$

where AD is the *average deviation*, $\Sigma|x|$ is the sum of *individual deviations* from the mean of the population, and N is the size of the population. The distance of any score from the mean $(X - \mu)$ is symbolized by a lowercase x^* and referred to as a *deviation score* (or sometimes simply *deviation*). The vertical bars in $|x|$ mean "without regard to sign."

The average deviation is sometimes used, but it is not common enough to be discussed here strictly for its own sake. Rather, I have included it because it is a direct expression of the basic idea that underlies the most used of all measures of variability. That idea is the concept of an *average of individual deviations* from the mean; that measure, however, is the *standard deviation,* not the average deviation.

Now compare the little-used statistic that we have been discussing to the standard deviation (which in many applications is the accepted measure of variability). Here again is Formula (4-1) for the average deviation:

$$AD = \frac{\Sigma |x|}{N}$$

and here is the formula for the standard deviation:

$$\sigma = \sqrt{\frac{\Sigma x^2_{\text{pop.}}}{N}} \qquad (4\text{-}2)$$

where σ is the standard deviation of the population, $\Sigma x^2_{\text{pop.}}$ is the sum of the squared deviations from the population mean, and N is the size of the population. The two formulas are very similar, aren't they? In fact, they are exactly alike except that for the standard deviation we take a mean of *squared* deviation scores; then we

---

*Some authors—in fact, most of them—use the *formula* for a deviation score as the *symbol* for it that appears in other formulas. For example, the formula for the average deviation would be

$$AD = \frac{\Sigma |X - \mu|}{N}$$

That notation is rather cumbersome, as you can see, so we'll be using x to stand for $(X - \mu)$—and, in the case of a sample, for $(X - \bar{X})$. Wherever it is not immediately clear from context which of these deviations is symbolized by x, be assured I shall make it so.

take the *square root* of that mean. But don't worry about the differences between Formulas (4-1) and (4-2). The important thing right now is their similarity: The standard deviation is a kind of average of individual deviations from the mean of the distribution.

In Chapter 3 you learned that the defining equation for a sample mean is essentially the same as the one that defines the population mean. The same is true of sample and population standard deviations, as you will see immediately when you compare Formula (4-2) with this one:

$$S = \sqrt{\frac{\Sigma x^2_{\text{sample}}}{n}} \qquad (4\text{-}3)$$

where S is the standard deviation of a sample, Σx^2 is the sum of the squared deviations from the sample mean, and n is the size of the sample.

As mentioned earlier, the standard deviation differs from the average deviation in that we *square* each deviation (which eliminates negative signs) and then find the square root of their mean. That mean (the mean of the squared deviations) is an entity in its own right *before* we take its square root: It is known as the *variance* of the distribution:

$$\text{Variance} = S^2 = \frac{\Sigma x^2}{n} \qquad (4\text{-}4)$$

where S^2 is the variance of a sample, Σx^2 is the sum of the squared deviations from the sample mean, and n is the size of the sample.

Let us take a moment to sum up: In this and the previous chapter we have examined a sequence of concepts that share a common structure. It is important for you to be aware of that structure, not only because it will help you to understand its manifestations in concepts developed earlier but also because it will appear in other concepts later. The best way to discern the structure is to place all the items of the series in close proximity to one another. Here are the four concepts in order in which they were presented:

$$\text{Mean of the raw scores} = \frac{\Sigma X}{n}$$

$$\text{Average deviation} = \frac{\Sigma |x|}{n}$$

$$\text{Variance} = \frac{\Sigma x^2}{n}$$

$$\text{Standard deviation} = \sqrt{\frac{\Sigma x^2}{n}}$$

They are all *averages*. The first is an average of raw scores; the other three are averages of deviation scores. In every case, the average is a mean. In the case of variance, it is the mean of the *squared* deviation scores. Last is the standard deviation, which is the *square root* of the variance and therefore is expressed in linear (*not* squared) units.

Appropriate Applications

The standard deviation is to measures of variability what the mean is to measures of central tendency. Of them all, it carries the most information, for it is sensitive to the value of every score, and in a normal distribution it is an integral component of many other useful statistics. Among those are the product-moment coefficient of correlation and the various standard errors, both of which will be discussed later. In scientific work, at least, the standard deviation clearly is the prevalent measure of dispersion.

As in Chapter 3, a calculation box (Box 4-1) provides you with an opportunity to try your hand at actually crunching some numbers. But here again, and for the same reason as before, the operations in the box are specified by the defining equation, not by a specialized calculating formula.

The Interquartile Range (IQR)

A much easier statistic to comprehend (and to compute) is the *interquartile range* (IQR). Look at Figure 4-4. What you see is a negatively skewed distribution cut into four equal parts. At the upper end of each of those quarters is a point on the baseline called a *quartile* (Q)—the first quartile (Q_1) above the lowest quarter, the second (Q_2) above the lowest two quarters,* and the third (Q_3) above the lowest three quarters. The point just above the highest score in the distribution would logically be Q_4, but that expression is seldom if ever used.

*What other familiar statistic has essentially the same definition as the second quartile? If you are not sure, turn back to page 29.

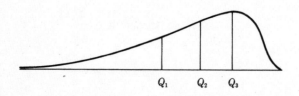

Q_1 Q_2 Q_3

Figure 4-4   Negatively skewed distribution with area divided into quarters.

Box 4-1 Calculation of Standard Deviation
(see Figure 4-2, verbal test)

(1) Class interval	(2) Midpoint X_m	(3) Frequency f	(4) $X_m - \overline{X}$ x	(5) $(X_m - \overline{X})^2$ x^2	(6) $f(X_m - \overline{X})^2$ fx^2
33–37	35	2	15	225	450
28–32	30	13	10	100	1300
23–27	25	32	5	25	800
18–22	20	106	0	0	0
13–17	15	32	−5	25	800
8–12	10	13	−10	100	1300
3–7	5	2	−15	225	450
		$\Sigma = 200$			$\Sigma = 5100$

Column 1: Class intervals. See pages 13 and 17–22.

Column 2: Midpoints of class intervals. Remember when data are grouped into class intervals, all individuals (X) within each interval are treated as though they were at the midpoint of that interval (X_m).

Column 3: Frequency (f) of scores in each class interval.

Column 4: Deviation scores (x), which equal the differences between the midpoints (X_m) and the mean of the distribution $(\overline{X})$. In this case, $\overline{X} = 20$.

Column 5: Squares (x^2) of the deviation scores listed in column 4.

Column 6: Product of the squared deviation scores and the frequency of scores in the interval (fx^2).

$$S = \sqrt{\frac{\Sigma x^2}{n}}$$

The sum of the frequencies listed in column 3 is n, and the sum of column 6 is Σx^2.

$$S = \sqrt{\frac{5100}{200}} = 5.01$$

Essence of the Concept

The *interquartile range,* like any measure of variability, is an interval. In this case the interval extends from Q_1, to Q_3. The calculation of the interquartile range is, therefore, extremely easy: You simply find the difference between Q_1 and Q_3. That's it.[1]

The IQR, like the median, is insensitive to the precise values of most of the scores in a distribution, so when you use it instead of the standard deviation you discard information. But if the distribution is skewed, you can retain some important information (namely, the direction of the skew) by reporting not only the *difference* between Q_1 and Q_3, but their exact *locations* as well, along with that of Q_2. (If $Q_2 - Q_1$ is longer than $Q_3 - Q_2$, the skew is negative, as in Figure 4-4; if $Q_3 - Q_2$ is the longer, the skew is positive.)

Appropriate Applications

The interquartile range is to measures of variability what the median is to measures of central tendency. Although insensitive to the exact values of many of the scores in a distribution, it is preferred to the standard deviation in the same kinds of situations in which the median is preferred to the mean—namely, when distributions are radically asymmetrical. It is the perfect companion to the median wherever the latter is properly applied.

A report of the incomes of a university faculty, for example, might be skewed by the high salaries and consulting fees of the college of business. If faculty incomes *outside the college of business* are distributed symmetrically, and you want to include the business faculty in a single distribution of all faculty incomes, the higher incomes in that college give the total distribution a marked positive skew, making the mean and standard deviation difficult to interpret. The median and interquartile range would be better in those circumstances than the mean and standard deviation, although again the best approach might be to treat the college of business separately from the general faculty.

The Range

There is another measure of variability—one that probably gets more attention than it deserves, both in makeshift analyses and in this book. It is given attention in makeshift analyses because it is so easy to compute, and much of the space assigned to it in this book is occupied by an exposure of its fundamental weakness and an attempt to prepare you for possible differences in its interpretation.

Essence of the Concept

Sometimes we are interested specifically in the most extreme cases in a sample; on these occasions we may report its *total range*—or simply *range*—which is the distance from the lowest score to the highest. As you can see, it is extremely easy to calculate.[2] That is about its only virtue, however.

In fact, the most important feature of the range is a weakness—its extreme instability. Note that the range in Figure 3-1 is 8. Now turn to Figure 3-2. It is the same as Figure 3-1 with the exception that two scores have been moved away from the main group. Note the effect on the range. (Instead of 8, it is now 35.5 − 1.5, or 34!) Note, too, that not even *two* extremely high scores were necessary to have that effect; one would have done exactly the same thing. The fact is that the range is determined by two, and only two, scores in any distribution: the lowest and the highest. (Compare that to the standard deviation, which is sensitive to *all* scores.) That is why the range is so easy to compute. That is also why it is so unreliable.

Appropriate Applications

We have seen that the standard deviation is analogous to the mean and is its proper companion, and we have noted a similar relationship of the interquartile range to the median. The relationship of the range to the mode is not quite as neat, but there are similarities that should help you to remember both.

One similarity is that each is the *quickest* index of its kind to calculate. Another closely related similarity is that each is *less stable* than alternative indices (although the range usually is much the worse in that respect because of its total dependence on only two scores). Finally, both the mode and the range are used, more often than are other statistics, to answer questions related directly and very simply to their definitions. For the mode, the question is "Which is the typical (most frequent) case?" For the range, it is "How much of the scale must be used to represent the distribution?"

SUMMARY

Two important ways of describing a sample are by its central tendency or average, which indicates the general level (magnitude) of the scores, and by its variability, which tells the extent to which individual scores deviate from that average. This chapter has been about the latter type of description—measures of *variability*.

You have been invited to use your previously developed understanding of measures of central tendency as an anchor for the new concepts; the latter are presented as analogs of the former. Specifically, the standard deviation is roughly analogous to the mean and is its companion statistic, the interquartile range goes with the median, and the range is similar to the mode.

The *standard deviation* is a kind of average of individual deviations from the mean of a distribution. The mean of the squared deviations is called the *variance,* and the standard deviation is the square root of the variance. Of all the

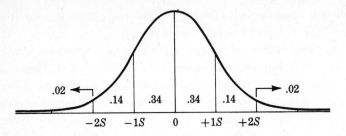

Figure 4-5   Normal distribution with baseline divided into standard deviations.

measures of variability, the standard deviation carries the most information. In a normal distribution it also lends itself to the computation of many more advanced measures.

The *interquartile range* is the distance from Q_1 to Q_3. It embodies less information than the standard deviation but is preferred to it whenever the distribution is markedly asymmetrical, and it is the perfect companion to the median.

The *range* is the distance from the lowest score to the highest, is completely determined by those two scores, and thus carries little information. Like the mode, it is both easier to compute and less stable than any other measure of its kind. It is, of course, the one statistic that gives information specifically about the distance between the highest and lowest scores in the sample.

We have seen that in a normal distribution, the mean, the median, and the mode are all at the same place. When we make a similar graphical comparison of measures of variability, the situation is not quite as simple.

In Figure 4-5, the baseline is divided into equal units—namely, standard deviations—and the areas subtended by them (.02, .14, .34, etc.) are unequal. The number above the line indicates the proportion of the total area that is subtended by each segment. The proportions have been rounded off because they are easier to remember in that form, and they are precise enough for our present purposes in any case. (More exact proportions are given in Figure 5-5, page 57.)

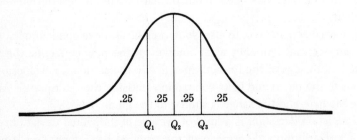

Figure 4-6   Normal distribution with area divided into quarters.

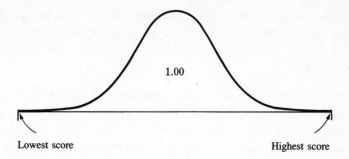

Figure 4-7 Normal distribution showing total range on baseline.

In Figure 4-6, the baseline is divided into *un*equal segments, and it is the parts of the area that are equal. And, of course, the range subtends the entire distribution, as shown in Figure 4-7.

Sample Applications

EDUCATION

You are appointed to a committee of elementary school teachers and administrators. The committee decides to make the teaching of reading a top priority for the next two-year period. You get permission to use the annual in-service training fund for purchasing a reading program in which all elementary teachers will participate. After an extensive search, you narrow the field to two programs that have been used and evaluated on usefulness and practicality by a large number of elementary school teachers. The means of their ratings of the two programs are approximately the same. Is there any other statistic that might help you make your decision?

POLITICAL SCIENCE

You are interested in the incidence of military coups in Latin America. Specifically, you want to know whether most Latin American countries have experienced a number of coups that is close to the mean number for the region. After you have gathered your data, how do you obtain that information?

PSYCHOLOGY

You are one of a five-person team of observers sent into a home to evaluate the degree of aggressiveness among family members over a one-week period. The observers indicate that, on the average, 8 aggressive acts occur per day. But there is also some disagreement. How might you quantify that disagreement?

SOCIAL WORK

You are the new director of a community fund-raising organization. The member agencies have widely varying needs, but you suspect that recently the board has been shirking its duty to investigate those needs. Specifically, you suspect that recent allocations have not been sufficiently differentiated. How might you document your case?

SOCIOLOGY

In order to make some decisions about the construction of family dwellings in your state, a construction firm asks you to report on the variability of family size. You have access to all of the state's data on family size. What measure of dispersion do you report?

5

Interpreting Individual Measures

Measurement has been defined as "rules for assigning numbers to objects to represent quantities of attributes."[1] The importance of *rules* lies primarily in the fact that the resulting assignment of numbers nearly always functions as a *communication*. Unless the receiver of a communication knows the rules by which the sender has made that assignment, however, the receiver will be unsure of its meaning. Many such rules are intuitively obvious. If you tell me that you have measured the distance from point *A* to point *B*, I shall assume that the rule you followed was to lay a ruler across the two points and note how many units (inches, for example) lie between them.

But most rules are not so obvious. What rule do you follow if you want to tell me the size of a particular circle? Do you lay your ruler across the center and the circumference and report the number of units between the two? If so, you had better label the resulting number as a "radius," so that I shall know exactly what you did to get the measurement. There are, of course, two other labels—"diameter" and "circumference"—that you could use and that would specify two rather different operations.

Those labels actually indicate different operations for measuring a circle, but because the operations have become standardized over the years, descriptions are no longer necessary to communicate precisely—the mere labels suffice. In some educational and psychological measurements, a similar standardization has occurred—though it is not as complete because the operations are much more complex. "Level of intelligence," for example, can be represented by a number, but the number is rendered more meaningful by a specification of the particular standardized test from which the number came, and even then the specificity of the operations performed falls far short of that indicated by the labels used in the measurement of circles.

47

Other measurements are not standardized at all. If we want to do a rat study in which one of the variables is "drive level," we have to figure out a way to measure that level; then, after doing so and running our experiment, we are obliged to describe in some detail the operations that resulted in whatever scores (measures) we have to report. It is in situations like this one that the need for specifying rules is most apparent; nevertheless, the rules are there, even when, as in the measurement of a circle, they are not explicitly stated. Were it not so, there could be no communication.

Our definition says that the rules we have been discussing are "for assigning numbers to objects." A number does not represent an object as such, but rather the quantity of some *attribute* of the object. How smart is this boy? How tall is that toy soldier? How long is the segment $A-B$ on the surface of this burial site? The numbers that we assign to the objects represent the magnitudes of their attributes.

The definition of measurement quoted above is adequate as far as it goes; it is a good general definition. But for our particular purpose, one further point needs to be made: In many measurement problems, scores can be legitimately interpreted only as *individual differences*. Indeed, individual differences may have an important effect on measurement even when the person doing the measurement is unaware of it. For example, Miss Jones has been teaching mathematics for 20 years. She tells all her students that they will be graded "on absolute standards," meaning standards inherent in the subject matter and independent of student performance in that subject matter; Miss Jones is openly contemptuous of "grading on the curve." * Indeed, we must acknowledge that if there is any discipline in which absolute standards should be used, mathematics surely must be it. But even in mathematics, when Miss Jones says that she is grading her students on absolute standards, there is reason to suspect that her statement is not entirely correct. She may require that first-year work be mastered before she passes a student to the second year, but she must have some idea, from either her own experience or that of her mentors and colleagues, what can reasonably be asked of first-year students. Her absolute standards turn out to be less absolute than she thought.

This example is from education. I am not saying that it is never possible to make an educational measurement of any one student without taking into account many other students on the same dimension. What I am saying is that in most applications, the more those other measurements are utilized in the interpretation of an individual score, the more sophisticated the interpretation will be.

---

*If Miss Jones had completed her teacher training 10 years later, she might be talking about "criterion-referenced" versus "norm-referenced" scoring instead of "absolute standards" versus "grading on the curve," but the different terms refer to very similar concepts.

Consider the case of Joseph O. Cawledge. JOC's high school grades were pretty bad, but then he was a full-time athlete, and the rest of the school's program never interested him very much. How well might he perform scholastically if he were motivated to do so? To find out, we give him Central University's scholastic aptitude test.

You will recall that the CU test is divided into two parts—a numerical test and a verbal test. Joe Cawledge takes them both because they are both required of all entering freshmen. He gets a score of 60 on the numerical test and a score of 30 on the verbal. Now, what do we know about Joe?

It *looks* as though he is twice as good with numbers as he is with words. But mightn't that appearance be deceiving? Turn to page 37 and place each of Joe's two scores within the appropriate distribution in Figure 4-2. (Insert a bookmark there, for we shall be using Figure 4-2 repeatedly for the next few moments.) In these as in all interval measurements, two questions must be answered about each scale: (1) What is the scale's *point of reference,* and (2) what is its *unit of measurement?* Answers to these questions will make it possible for us to answer two parallel questions about any individual score: (1) Is it *above or below* the reference point, and (2) *how far* is it from that point?

It is customary to use the mean of some well-defined "standardization" or "norm" group as the standard reference point. You can see immediately that each of Joe's scores is above that point on the appropriate scale. But how far above? "Ten points each. They are both the same distance above their respective means." Is that what you say? Look again. One score (the numerical) is in the populous middle part of its distribution, but the other (the verbal) is way out in the tail of its distribution. On second thought, doesn't it seem to you that the latter must be a much higher score than the former?

Yes, but how much higher? Once again we find that drawings are excellent aids to comprehending basic principles, but in many applications we need a numerical index. We need a single number that will tell us how far away from the mean any given score is and in what direction.

Standard Scores: The z Scale

The answer to the question of how much higher Joe's verbal score is than his numerical must be in terms that make it possible to *compare* an individual's scores on two different scales. To put it another way, we must find a way to put both scores onto a single scale—a *standard* scale, if you will. That probably doesn't mean much to you right now, but in a moment it will.

We have already found a common reference point for both distributions, haven't we? Every distribution has a mean; therefore, the mean can be used as the reference point for the standard scale that we are about to develop. Every score that falls precisely on the mean of its own distribution will be given a score of 0, because all counting of units will start from there.

Now that we have a common reference point, all that remains is to find a common *unit* for our new scale. As Figure 4-2 demonstrates, it must be a unit that can be used to *compensate for the variability* of a distribution; in order to do that, it must *be* a measure of variability. If we were to divide Joe's numerical deviation score by a large number and his verbal score by a small one (look back at Figure 4-2), wouldn't we have a more realistic index of his position in each distribution for purposes of comparison?

Because a measure of variability would be large for the numerical and small for the verbal distribution, that is how we shall accomplish our purpose of converting both scores to a common scale. If you had a measurement stated originally in inches and you wanted to convert it to feet, how would you proceed? You would divide by the number of inches in a foot, wouldn't you? Well, you do the same thing here: You divide by the number of raw-score units in a *standard deviation* (the standard deviation of that particular distribution). The result gives you the number of standard deviations in the interval being measured. The formula looks like this:

$$z = \frac{x}{S} \tag{5-1}$$

where z is a standard score, x is a deviation score, and S is the standard deviation of the distribution in which x occurs. (In this case that would be, for each test, the "norm group" mentioned earlier.)

Let's see how it works in Joe's case. If the standard deviation of the numerical distribution is 15, and Joe's score of 60 is 10 raw-score points above the mean, his deviation score is 10 and his standard score is $\frac{10}{15}$, or 0.67. If the standard deviation of the verbal distribution is 5, and Joe's score of 30 is again 10 points above the mean, his deviation score is 10, but his standard score is $\frac{10}{5}$, or 2.00. Quite a difference, isn't it? Figure 5-1 shows where Joe's two scores would be in a distribution of standard scores from both tests.

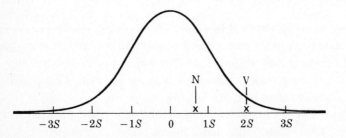

Figure 5-1   Distribution of standard scores from two different tests. N = numerical; V = verbal.

We have accomplished our purpose: We have moved scores from two different scales onto a single standard scale. When each was expressed in terms of its own scale, the two could not be compared; now they can be.

You perform the same sort of operation intuitively all the time: A "small elephant" is much larger than a "small mouse," and the size difference between a small elephant and a large one is much greater than the difference between a small and a large mouse. That is insightful thinking; what standard scales can add to it is precision.

Other Standard Scores

Effective though it is, our standard scale still has some shortcomings in actual practice. For one thing, the negative numbers that can result are a nuisance. (If instead of 60 and 30 on the numerical and verbal tests, Joe had made a raw score of only 20 on each, his standard scores would have been -2 and 0, respectively.) Another shortcoming is that the scale unit (1 standard deviation) can be awkwardly large, thus necessitating the use of (decimal) fractions to achieve the desired precision.

Fortunately, however, neither of those shortcomings is difficult to overcome. To get rid of the negative numbers, we have but to add a constant to every z score; if that constant were 5, the mean score would be $0 + 5 = 5$, 1 standard deviation below the mean would be 4, 2 above it would be 7, and so on (see line B in Figure 5-2). To reduce the size of the scale unit requires nothing more than dividing it into smaller parts, which of course makes the *number* of parts larger. If we want our new units to be $\frac{1}{10}$ the size of the original, there will be 10 of the

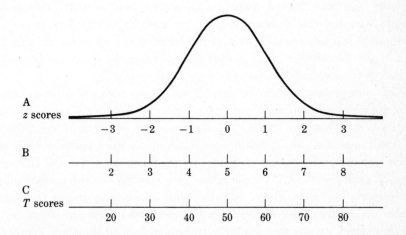

Figure 5-2 Evolution of a derived standard scale with a mean of 50 and a standard deviation of 10.

new units to every standard deviation. Combining those maneuvers (first adding 5 to every z score and then multiplying by 10) produces the scale of T scores that you see in line C of Figure 5-2.[2] The result is often referred to as a derived standard scale and its scores as derived scores—or sometimes "scaled scores."

Other derived scales are constructed in much the same way. In the scale for the College Boards (the CEEB tests described on page 56), 5 again is added, but this time 100 is the multiplier; the resulting distribution has a mean of 500 and a standard deviation of 100. The Army General Classification Test mean ("AGCT scores," also described on page 56) is arbitrarily set at 100 and its standard deviation at 20. In every case, what we have is but a modification of what is still basically a standard deviation scale.

Centile (or Percentile) Scores

Standard, or scaled, scores can be used to compare an individual's scores on two different tests or to trace with one test his progress over time. Whatever else it may do, every such score compares the individual's performance to that of a standardization group. But the size of that group is commonly monstrous and its composition heterogeneous. Often what is most needed is a comparison with a group that is less cumbersome and more precisely defined.

Take Joe Cawledge's numerical performance, for example (page 49). Imagine now that the test, far from being homemade by Central University as I have asked you to think of it until now, is only being borrowed by the university as a part of a national testing program. The test has already been *standardized*— that is, it has been administered to a very large number of college students, measures of central tendency and variability have been computed, scaled scores (say, T scores) have been derived, and *norm tables* have been prepared that make it possible to convert quickly any individual raw score to other kinds of scores. (You will recall that a raw score by itself is meaningless.) Imagine further that Joe's score on the numerical test is 1 standard deviation above the national mean (it was only $\frac{2}{3}$ of a standard deviation above the local mean).

That is important information in itself, and it also makes possible a comparison of Joe's present performance with his performance on the same test months or years later. What this information does *not* do, however, is to compare him with the competition at CU; more specifically, it does not compare his performance with the performances of other entering freshmen at Central University. However, each year the CU people gather their own data on the performance of their entering freshmen. Those data then appear as local norms in the form of *centiles*—or, as they are more often called, *percentiles*. Actually, of course, the data themselves are raw scores; they must be transformed somehow in order for the information to be used in this new way.

What is this new way? It is in fact very simple. Do you remember the concept of the quartile (page 40)? It is the number of quarters of the distribution that lie below the score being reported. If the point above a quarter is a quartile, how might we refer to the point above a *tenth* of the scores in a sample? *Decile* would be the parallel term. And what if we were to divide the sample into 100 equal parts? The parallel term there would be *centile*, wouldn't it? A centile is the number of *hundredths* of the distribution that lie below the score being reported. The term *percentile* is technical slang for *centile*.

Now back to Joe. His numerical raw score was 60, which happens by coincidence to be precisely 1 standard deviation above the national mean and converts to a *T* score that is also 60. That conversion has been done by a computer at national headquarters, and neither Joe nor his counselor ever sees the raw scores. What they do see is a standard score of 60 and a norms table that allows them to convert that score into a percentile in any of several populations. (Samples from those populations are called *norm groups,* and the process of gathering data from them is often referred to as the *standardization* of the test.) Table 5-1 shows how the norms might look for the numerical test. The table is represented graphically in Figure 5-3.

The important thing to notice is that Joe has not one score but many. If it is true that his performance is meaningless until it has been compared with other performances, it is also true that its meaning is enhanced by a precise definition of the group with which he is being compared at any given time. Once that definition has been made, individuals who fit it can be tested and the distribution of their scores divided into hundredths. The result is a set of *norms* like that in Table 5-1.

Now our Joe can see how his performance compares not only with the performances of the entering American college freshmen on whom the numerical test was originally standardized, but also with those of any other group that has been tested. His score is better than 84 percent of entering freshmen, 79 percent of students completing their freshman year, 71 percent of completing sophomores, 63 percent of completing juniors, and 55 percent of graduating seniors. Those norm groups are students from all kinds of programs, including art, music, literature, and so forth, in which little numerical work is done. If Joe were thinking of entering an engineering curriculum, it would be appropriate to use norms derived from the testing of engineering students. Meanwhile, we should suspect that his score, compared with scores of engineering freshmen, might be well below the median and that among graduating engineering seniors it might even be near the bottom.

The small curves in Figure 5-3 have been artificially equated in size and shape. You might expect the "completing freshmen" group to be closer to the size of the "entering freshmen" group than the drawing might lead you to believe, and you would be right. You might also expect that the shapes of the distributions

TABLE 5-1   Norms for numerical test

Standard score	Percentile				
	Entering freshman	Freshman	Sophomore	Junior	Senior
88					
86					
84					
82					
80					99
78				99	98
76			99	98	97
74		99	98	97	95
72	99	98	97	95	92
70	98	96	95	92	88
68	96	94	92	88	83
66	95	92	88	84	78
64	92	88	84	78	71
62	88	84	78	71	63
60	84	79	71	63	55
58	79	72	63	55	47
56	73	64	56	48	39
54	66	57	48	39	31
52	58	49	40	31	24
50	50	41	32	24	18
48	42	33	25	18	13
46	34	26	19	13	9
44	27	20	13	9	6
42	21	14	10	6	4
40	16	10	7	4	2
38	12	7	4	2	1
36	8	5	3	1	
34	5	3	2		
32	4	2	1		
30	2	1			
28	1				
26					
24					
22					
20					

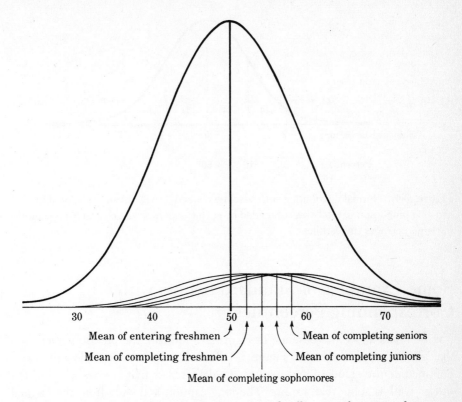

Mean of entering freshmen

Mean of completing freshmen

Mean of completing sophomores

Mean of completing juniors

Mean of completing seniors

Figure 5-3   Distribution of four survivor groups artificially equated in size and superimposed upon original freshmen distribution.

would change systematically from entering freshman to graduating senior; after all, it is the less able whose elimination accounts for the rising averages. But there you would be wrong. In practice, it turns out that distributions of more select groups tend to be very nearly normal.

One thing more. In Chapter 4 (Figure 4-5) we identified the proportions of the total sample that, in a normal distribution, are subtended by successive standard deviation units marked off from the mean. Percentiles might be thought of as "cumulative percentages." Figure 5-4 reproduces Figure 4-5 and adds the percentile that corresponds to each of the standard scores.

You should be able to reproduce this diagram from memory. However, when I say "from memory," I do not mean rote memory. You need memorize only two numbers: 34 and 14. Once you have placed those properly, the rest are determined. Try it.

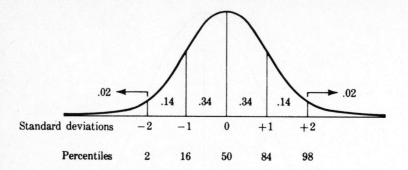

Standard deviations   −2   −1   0   +1   +2

Percentiles   2   16   50   84   98

Figure 5-4 Normal distribution with baseline divided into standard deviation units, showing proportion of total area subtended by each standard deviation and listing cumulative percentages (percentiles).

Some Examples of Standard Scores with Corresponding Percentiles

The numbers above the baseline in Figure 5-5 indicate the precise percentages that are only approximated in Figure 5-4. The alternative scales below the baseline include the standard and derived standard scales that are described in the text as well as some that are not. The long caption is quoted from the original Psychological Corporation bulletin.

Figure 5-5 lists four tests to which you have not been introduced:

- *CEEB* stands for College Entrance Examination Board. It publishes the SAT (Scholastic Aptitude Test), which is actually two tests—Verbal and Math. The scale shown applies to either of those two, but the composite is their sum; so if a student's performance is precisely average on each of the two tests, his composite will be not 500 but 1000.

- *AGCT* means Army General Classification Test. It is a group test of general intelligence and one component in a battery of instruments that help the U.S. Army assign recruits to specialized training after their basic training has been completed.

- *Stanines* refer not to any particular test but to a derived standard scale that could be used to report the results of *any* measuring operation. (Pilots, bombardiers, and navigators in World War II were selected partly on the basis of their stanines on a battery of psychomotor tests.) Stanines can be conceived as a compromise between z units, which everyone agrees are too large for many applications, and *T* units, which some users believe are too small.

- *Wechsler Scales* report performance on a widely used individually administered intelligence test (either the Wechsler Adult Intelligence Scale or the

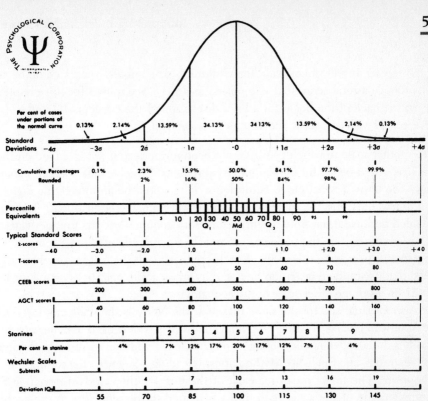

Figure 5-5 The normal curve, percentiles, and standard scores. Distributions on many standardized educational and psychological tests approximate the form of the normal curve shown at the top of this chart. Below it are some of the systems that have been developed to facilitate the interpretation of scores by converting them into numbers which indicate the examinee's relative status in a group.

The zero (0) at the center of the baseline shows the location of the mean (average) raw score on a test, and the symbol σ (sigma) marks off the scale of raw scores in *standard deviation* units.

The cumulative percentages are the basis of the *percentile equivalent* scale.

Several systems are based on the standard deviation unit. Among these *standard score* scales, the z score, the *T* score, and the stanine are general systems which have been applied to a variety of tests. The others are special variants used in connection with tests of the *College Entrance Examination Board,* the World War II *Army General Classification Test,* and the *Wechsler Intelligence Scales.*

Tables of norms, whether in percentile or standard score form, have meaning only with reference to a specified test applied to a specified population. The chart does not permit one to conclude, for instance, that a percentile rank of 84 on one test necessarily is equivalent to a z score of +1.0 on another; this is true only when each test yields essentially a normal distribution of scores and when both scales are based on identical or very similar groups of people.

The scales on this chart are discussed in greater detail in *Test Service Bulletin No. 48,* which also includes the chart itself. . . . Copies of this bulletin are available from The Psychological Corporation, 304 East 45th Street, New York, N.Y. 10017. [Psychological Corporation, "Methods of Expressing Test Scores," *Test Service Bulletin No. 48,* September 1954.]

Wechsler Intelligence Scale for Children). The subtest scores combine to produce a total score that is reported as an IQ. See note 3 for this chapter, in the back of the book, for a brief discussion of the concepts *IQ* and—especially with respect to the Wechsler—*derived IQ*.

Another scale to which you have not been introduced is *not* included on the list. It is called *NCE,* for *normal curve equivalents*. (To elucidate the full meaning of the term "normal curve equivalents" here would be an interesting digression but would consume more space than could probably be justified.) It is not included because it has only recently come into common use. Its unique feature is that a score of 1 on the scale corresponds to a percentile of 1 and a score of 99 corresponds to a percentile of 99. To achieve that result the test makers have decreed that its standard deviation be not 1, as on a z scale; not 10, as on a T scale; but 21.06, as on no other scale of my acquaintance.

Also missing, and for the same reason as the NCE, is the scale employed by the American College Testing Program. The ACT is widely used in collegiate admissions and academic counseling in place of or in addition to the SAT. Scores are reported on English, Mathematics, Reading, Science Reasoning, and Composite. The target mean for each of these is 18, the standard deviation 5. The magnitude of the scale score units is related to the *reliability* of the score. For any of the four tests the scale segment *between 2 points below and 2 points above* the score has an unique property: There is a probability of approximately 68 percent that the *true* score lies within that interval. For the Composite score the corresponding interval is only half that size (from *1* point below to *1* point above), because including as it does four tests instead of one, the Composite score is more reliable than any one of its component scores. (This kind of reliability is discussed in Chapter 8.)

Age and Grade Norms

A percentile tells us where an individual is in relation to a specified norm group. Another way to interpret a scaled score (or sometimes even a raw score) is to find that group which the individual's score most nearly resembles. For example, each comparison group can be taken entirely from a particular chronological age group or from a particular grade in school.

By far the best known of age norms is the *mental age,* which used to be a prerequisite to the derivation of an intelligence quotient (IQ). An individual child's score is compared with the average (usually median) scores of many chronological age groups; the age of the group whose average is closest to his determines his mental age. For example, if six-year-old Cathy does as well as an *average* six-year-old on an intelligence test, her mental age is 6; but if she does as well as the average eight-year-old, her mental age is 8, not 6.[3]

The comparison group usually used in interpreting scholastic achievement tests is the *school grade* rather than the age group. If a child's score is nearest to that made by the average entering sixth-grader, his *grade level* is said to be 6.1, meaning "the first month of the sixth grade." If a child's score is equal to the average pupil in the middle of the sixth year of school, his grade level is 6.5; if his performance is most similar to that of an average seventh-grader in the third month, his grade level is 7.3; and so on.

SUMMARY

Measurements in the social sciences are nearly always of *individual differences*. The report of a score can seldom be understood out of context; that is, it can be understood only in relation to other scores.

A *standard score* results from placing a score from any test onto a scale common to all tests. That is accomplished by dividing each individual deviation score by the standard deviation of its own distribution. Other standard scores are *derived* from this one basic unit.

A *centile* (or *percentile*) shows where an individual stands in relation to a specified norm group. It does so by reporting just what percentage of that distribution falls below the individual's score.

An *age* or *grade level* is the identification of the norm group whose average score is most similar to that of the individual tested.

The essence of each of those procedures is *comparison* of an individual to a defined group. A standard scale divides the *baseline* of a distribution into equal parts, and percentiles divide the *area* under the curve (the total sample) into equal parts. Age or grade norms are the average scores of various age or grade groups. Thus, all of these measures are different, but each can be used to compare a new subject's score with the many other scores that have preceded it.

Sample Applications

EDUCATION

A fourth-grade student in your school has been having considerable difficulty in several subjects. His teacher has tried various approaches, but although the student can read words at grade level, he does not seem to remember what he reads, and although he can complete simple addition and subtraction problems, he cannot complete problems that involve several steps. The teacher refers the student to you, the school psychologist, and you give the student a battery of tests to determine his specific strengths and weaknesses. After the tests have been scored, what should you do before making your report?

POLITICAL SCIENCE

Many of the concepts in political theories are so complex that they cannot be expressed in terms of simple indices. As a result, researchers often create complex indices by summing several separately derived scores, each of which represents a different dimension of the concept. Imagine, for example, that you want to measure civil strife. A simple index (such as the frequency of riots) would not suffice for measuring such a complex concept, but adding the scores on labor-hours lost per strike, assassinations per 1000 population, and other such indices derived through different measuring procedures would not work either, because each is measured in a unit different from that used in measuring each of the others. What can you do?

PSYCHOLOGY

June is a year-old infant, the product of a difficult pregnancy and premature delivery. At birth, the attending physician feared that June's development would be delayed or fragmented (high in some areas, low in others).

You are asked to evaluate the quality of June's overall development. You are concerned about her growth in three different areas: intellectual, social, and psychomotor. You administer separate tests corresponding to those areas and get three measures (scores). However, each of the three distributions has a different mean and standard deviation from the other two. How can you compare June's development in the three areas?

SOCIAL WORK

The state personnel office conducts an examination to establish a hiring register for social workers. It consists of a battery of tests concerning such functions as client assessment, client treatment, and community resources. Each test has its own mean and standard deviation, but all have been standardized on the same population. How can the personnel office inform you about your relative performance on the various parts of the examination?

SOCIOLOGY

A study of authoritarianism in the various departments of your university has just been completed, and a friend of yours has access to the results. You are about to enroll in a required English course, but because you yourself harbor authoritarian attitudes you want to postpone enrollment unless the course is offered this semester by a professor with attitudes similar to yours. What information do you request from your insider friend?

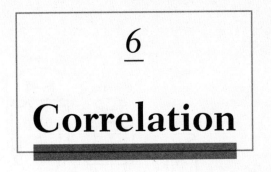

6

Correlation

There are many times, both in basic science and in professional practice, when we want to know the *relationship* between one thing and another. Indeed, all of science is concerned with such relationships, and without knowledge of them, professional practice could never check up on itself.

Take our scholastic aptitude test, for example (pages 13ff). If we merely assume that it is measuring what we want it to, then much money and even more time and effort may be spent in vain. If, on the other hand, we define the test's effectiveness in terms of its prediction of success in college, then we have a way of checking up on it. Once we discover how closely the scores are related to some criterion of success, we can judge the test's effectiveness.

What we need, then, is an index of relationship—a number that when low indicates a low degree, and when high a high degree of relationship between two variables. We need a *coefficient of correlation*.

Imagine now that the two variables in which we are interested are the height and the weight of a population of toy soldiers. Imagine further that all the soldiers are exactly the same *shape* and that they differ in size and therefore in weight. (This is, of course, an unnatural situation; I am using it because it gives me absolute freedom to manipulate both variables so as to present positive correlation in its purest form. Real examples are never so elegant.) Figure 6-1 pictures five soldiers. If we were actually doing the computation, we would need a much larger set than five, but for learning the concept, the smaller number is better.

Just by looking at the five soldiers, what would you say the relationship is between height and weight? You notice immediately that the short soldiers are light in weight and that the tall ones are heavy. How would you describe that relationship? Strong? Directly proportional? Perfect? If you were asked to place the strength of the relationship on a scale from .00 to 1.00, you would have to place it at 1.00, because it is perfect.[1]

A coefficient of correlation provides just such a scale—that is, a scale with limits of .00 and 1.00—except that it carries information not only about the

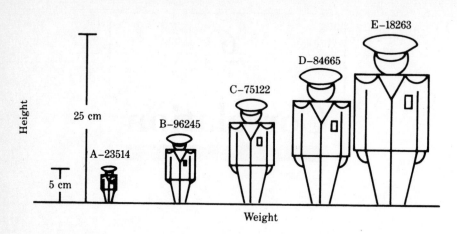

Figure 6-1 Sample of a population in which height and weight have a perfect positive correlation. (The label above each toy soldier is a serial number for identification.)

strength of the relationship but also about its *direction;* some correlations are positive and some are negative.

There are several kinds of coefficient. The easiest to comprehend is the rank-difference coefficient; the most useful (and most used) is the product-moment coefficient. We shall examine both.

The Rank-Difference Coefficient (ρ)

Near the beginning of Chapter 4, "Measures of Variability," I introduced you to a statistic that you will probably never see again. I did so because that statistic, the average deviation, was the best device available for building an understanding of the concept that lies behind the most used of all measures of variability, the standard deviation. What I am about to do with correlation is not quite as drastic as that, for the rank-difference coefficient is used quite often. But it appears in the literature much less often than the product-moment coefficient, and again I am presenting the less frequently used statistic first because it illustrates more directly the essential nature of the more common one.

The Essence of Correlation

Table 6-1 shows how the data would be ordered in preparation for computing a Spearman rank-difference coefficient of correlation on the toy soldiers in Figure 6-1.

In Table 6-1, the first three columns list each soldier's serial number, rank on the variable *height,* and rank on the variable *weight,* respectively. The fourth column shows the difference between the two rankings. If you draw straight

TABLE 6-1   Ordering of data for rank-difference correlation of heights and weights of toy soldiers in Figure 6-1

Identification of subject	Rank on variable X (height)	Rank on variable Y (weight)	Difference between ranks, D	Square of difference, D^2
E-18263	1	1	0	0
D-84665	2	2	0	0
C-75122	3	3	0	0
B-96245	4	4	0	0
A-23514	5	5	0	0
				$\Sigma = 0$

lines between equal ranks in the second and third columns, the lines will form a ladder pattern. That has been done in Table 6-1. (In a moment, we'll look at a similar table that forms a different pattern.) Notice, too, that *all the rank differences are zero*. Keep that in mind while looking at the formula for *rank-difference coefficient*, labeled ρ (rho):

$$\rho = 1 - \frac{6\Sigma D^2}{n(n^2 - 1)} \tag{6-1}$$

where ρ is Spearman's rank-difference coefficient, D is the difference between the two ranks of a single subject, and n is the number of subjects—that is, the number of *pairs* of measurements. We are not interested in the computation as such[2]; the formula is included here only to demonstrate that the rank differences are in the numerator of a fraction that is subtracted from 1.00. That means that the larger the rank differences, the smaller the ρ, down to a limit of .00 (after that, still larger differences contribute to rising *negative* coefficients, up to a limit of -1.00).

Any correlation coefficient carries information about two aspects of a relationship: its *strength*—measured on a scale from zero to unity—and its *direction*—indicated by the presence or absence of a minus sign. Consider again our example in Figure 6-1. I asked you to imagine that all the subjects were exactly the same shape, and I had attempted to draw them all the same shape. But my draftsmanship falls short of perfection (only slightly, mind you!), so if we were to cast those toy soldiers in molds made directly from my drawings, the relationship between height and weight would *not* be perfect, after all. You might expect a coefficient of correlation to be sensitive to that difference between perfection and near-perfection, but the rank-difference coefficient is not. Examine

Figure 6-1 again, and you will see that all the subjects clearly are ranked the same even when the imperfections of my drawing are recognized. I mention that here because later I want to show you an index (the product-moment coefficient) that *does* take those imperfections into account.

As for the negative sign that accompanies some coefficients, there is an important point to be made (or rather, emphasized, for it has already been made). *The size* (strength) *of a coefficient is entirely independent of its direction* (positive or negative). Although the formula for the rank-difference coefficient does not encourage it, we should think not of a single continuum from negative 1.00 through .00 to positive 1.00 but rather of two separate dimensions—one negative, one positive—each of which begins at .00 and ends at 1.00. A correlation of +1.00 is no bigger (stronger) than a correlation of −1.00.

Negative Correlation

What does it mean to say that two variables are *negatively* correlated? Let us look again at the variables of height and weight, but this time in a population quite different from the toy soldiers that we examined in Figure 6-1 and Table 6-1. Consider now a human population in which the shortest individuals are the heaviest and the tallest are the lightest. (Such an arrangement is beyond your experience, but not your imagination.) Figure 6-2 is a rendering of a sample of five soldiers drawn from such a population. Table 6-2 reveals the different pattern I promised a moment ago. If you look at the lines drawn between equivalent ranks, you will find that a star pattern emerges instead of the ladder that you see in Table 6-1.

If you sum the squared rank differences, you will find a marked contrast with the zero that comes from Table 6-1. (Remember that D^2 is in the

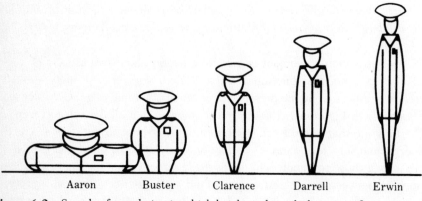

Figure 6-2   Sample of population in which height and weight have a perfect negative correlation.

TABLE 6-2   Ordering of data for computing rank-difference correlation of heights and weights of men in Figure 6-2

Identification of subject	Rank on variable X (height)	Rank on variable Y (weight)	Difference between ranks	Square of difference
Erwin	1	5	−4	16
Darrell	2	4	−2	4
Clarence	3	3	0	0
Buster	4	2	2	4
Aaron	5	1	4	16
				$\Sigma = 40$

numerator of a fraction that is *subtracted* from 1.00 to obtain the correlation co-efficient.) But my drawing is a considerably less adequate representation of a perfect relationship than was Figure 6-1. Does the ρ coefficient detect my inadequacy this time? No. Nothing will affect it until the *ranks* change. In Table 6-2, the correlation of height to weight is a full −1.00, even though the weights of Darrell and Erwin, for example, are perceptibly farther apart than their heights.*

In other words, information is lost when we convert interval measurements into rank orders.[3] That difficulty can be overcome, at considerable cost in computational labor, by using a different coefficient of correlation. The next section will aim at an understanding, with little computational labor, of that coefficient.

The Product-Moment Coefficient (*r*)

The ρ coefficient is included in this book primarily because it makes a better vehicle for teaching than the more difficult Pearson *product-moment coefficient*. You will see ρ reported from time to time in the literature, but usually it is used only as a relatively quick approximation of the "Pearson *r*," as the product-moment coefficient is often called, or in situations where the assumptions for *r* cannot be satisfied.[4]

---

*That difference may not be as perceptible to you as it is to me, so let me give you an exaggerated example. In the following diagram, the arrangement of subjects A, B, and C on variable X is quite different from that of the same individuals on variable Y, but their ranks are the same on both:

X:	A	B		C
Y:	A		B	C

The Meaning of "Product Moment"

To begin our discussion of this more sophisticated index, let us look again at those toy soldiers (Figure 6-1). Once again we construct a table, and indeed it is of the same general form as Table 6-1, but this time instead of ranks, we enter ordinary interval scores that also tell how far apart the subjects are on each variable (Table 6-3). (See Note 1, Chapter 10, on page 175.)

Don't be intimidated by that table. One reason it is so large is that information must be collected for the computation of two standard deviations. (The x^2 and y^2 columns are used exclusively for that.) The reason for including that computation should be clear, based on the discussion of standard scores in Chapter 5. The formula for r is:

$$r = \frac{\Sigma xy}{n S_x S_y} \qquad (6\text{-}2)$$

where r is the Pearson coefficient, Σxy is the sum of the cross products of deviations from the means of X and Y distributions, n is the number of such products (or of subjects or of pairs of observations), and S_x and S_y are the standard deviations of the two distributions.

The main idea behind this formula is that when X and Y scores are arranged in parallel, so to speak, the product of the corresponding deviation scores (xy) is maximally large. This is so because in such an arrangement the largest deviation scores—both positive and negative—occur together in the table and thus are multiplied together. Compare the "ordered" and "random" parts of Table 6-4 and you will see what I mean.

Close examination of the two parts will reveal that the scores are the same in both but that they are arranged differently. On the left, the arrangement is perfectly ordered, as the drawing of toy soldiers in Figure 6-1 would suggest; on

TABLE 6-3 Ordering of data for computing product-moment correlation of heights and weights of toy soldiers in Figure 6-1

Serial number	X (height in centimeters)	Y (weight in grams)	x	y	x^2	y^2	xy
E-18263	25	250					
D-84665	20	160					
C-75122	15	90					
B-96245	10	40					
A-23514	5	10					

TABLE 6-4   Effect of ordering on sum of cross products (Σxy)

	Ordered					Random			
X	Y	x	y	xy	X	Y	x	y	xy
25	250	10	140	1400	25	160	10	50	500
20	160	5	50	250	20	40	5	−70	−350
15	90	0	−20	0	15	10	0	−100	0
10	40	−5	−70	350	10	250	−5	140	−700
5	10	−10	−100	1000	5	90	−10	−20	200
Σ 75	550			3000	75	550			−350
$\overline{X}$ 15	110				15	110			

the right, the arrangement of the Y scores is random, which means that the relationship between X and Y is also random.

Now see what that arrangement does to the correlation coefficient. On the left, every large positive deviation score gets multiplied by another that is similarly large and also positive; whereas on the right, a large positive deviation may be neutralized by a low multiplier, and *any* positive deviation may be multiplied by one of an opposite sign, thereby yielding a negative product. The result is that with the random arrangement, the sum of the *cross products* (Σxy)—and hence the correlation coefficient—is always small. (In fact, it varies from *zero* only by chance!) In terms of the weightless beam concept introduced on page 26, you could say that the deviation scores on the two variables are often either close to the fulcrum or on opposite sides of it, whereas in a more substantial correlation, x and y scores for a given individual are consistently together—frequently far from the fulcrum—so that they can apply an extremely large torque, or moment, to the beam. Hence the term *product moment*—a reference to the product of the moments, which is largest when all the scores are perfectly ordered. Our defining equation for the Pearson product-moment coefficient is given in Formula (6-2). Box 6-1 shows how an r might be calculated using that formula.

The Scatterplot

There is another way of looking at correlation that gets directly at the fundamental idea. I am referring to the *scatterplot*. A scatterplot (or *scattergram*) is a two-dimensional approximation of a three-dimensional frequency distribution. It is a surface on which both the X and Y scores of each individual can be represented by a single point. If, for example, we were correlating scholastic aptitude test scores with first-year grade point averages and if all students' test scores were precisely proportional to their subsequent scholastic performance, the

Box 6-1 Calculation of Correlation Coefficient, r (data not discussed in text)

(1) Sugar intake X	(2) $X - \overline{X}$ x	(3) Time on task Y	(4) $Y - \overline{Y}$ y	(5) xy
45	15	22	−28	−420
43	13	18	−32	−416
42	12	50	0	0
40	10	64	14	140
37	7	56	6	42
36	6	44	−6	−36
35	5	41	−9	−45
30	0	59	9	0
25	−5	36	−14	70
24	−6	73	23	−138
23	−7	31	−19	133
20	−10	78	28	−280
18	−12	69	19	−228
17	−13	27	−23	299
15	−15	82	32	−480
				$\Sigma = -1359$

Column 1: Sugar intake as percentage of total carbohydrates ingested (X).

Column 2: Deviation scores (x) of sugar intake, which equal the differences between the individual scores (X) and the mean of the X scores ($\overline{X}$; see sample calculation on page 28). In this case, $\overline{X} = 30$.

Column 3: Time on task as percentage of time available (Y), used as an index of hyperactivity. (A *low* score indicates hyperactivity.)

Column 4: Deviation scores (y) of time on task, which equal the differences between the individual scores (Y) and the mean of the Y scores ($\overline{Y}$). In this case, $\overline{Y} = 50$.

Column 5: Product of deviation scores, x and y. Notice that a negative product occurs only where a negative x is paired with a positive y or vice versa.

$$r = \frac{\Sigma xy}{n S_x S_y}$$

The number of xy products is n. The sum of the column 5 is Σxy. $S_x = 10$ and $S_y = 20$. (See page 41 for a sample calculation of the standard deviation.)

$$r = \frac{-1359}{(15)(10)(20)} = -.45$$

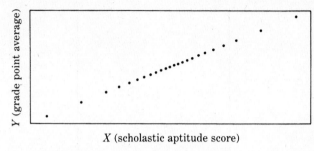

Figure 6-3   Scatterplot of perfect positive correlation.

scatterplot might look like Figure 6-3. Each student is represented by a single point (here a black dot) that locates the student on two dimensions, *X* and *Y*.

A third dimension is formed whenever two or more individuals occupy the same point on that two-dimensional surface. Because the third dimension is extremely difficult to represent on the flat pages of a book, I have, in places where individuals tend to pile up, separated them enough so that you can see all of them.

In the case represented by Figure 6-3, the students' scores on *X* and *Y* form a straight line (called a *regression line** because it embodies the mathematical regression of *Y* on *X*) with no scatter. If the scores were all in standard units (discussed in Chapter 5), the regression line in a perfect positive correlation (e.g., the zero scatter of Figure 6-3) would necessarily have a slope of 1.00, and any smaller correlation (e.g., the moderate scatter of Figure 6-4) would produce a slope smaller than 1.00. When scores are in standard units, the smaller the

---

*Regression in this context is a functional relation of correlated variables; in this application, the correlated variables are labeled *X* and *Y*. The *regression line* yields an approximation of the mean value of *Y* for any specified value of *X*: a line of best fit. For example, in Figure 6-11B, page 77, if *X* is 1.0, the mean value of *Y* is about 0.75; and if *X* is 2.0, the mean of *Y* is about 1.5. The slope of the line is 0.75. But in Figure 6-11C, if *X* is 1.0, the mean of *Y* is 0; if *X* is 2.0, the mean of *Y* is still 0. The slope of that line is 0.00.

The slope of a line in a coordinate plot like any one of those in Figure 6-11 is the ratio of the amount of change on the ordinate (*Y*) to the amount of change on the abscissa (*X*). If *Y* increases $\frac{1}{2}$ unit for every unit increase in *X*, the slope is 0.5. If while following a regression line with your pencil you find that you have to move 4 units on the *X*-axis for every 1 you cover on *Y*, the slope is 0.25—providing that the changes on both variables are positive. The same *degree* of slope could occur in a *negative* direction if *Y* were *dropping* $\frac{1}{4}$ unit with every 1-unit rise in *X*. All of this assumes that both *X* and *Y* have been converted to standard scores, which in Figures 6-3 through 6-7 they have not.

The relation between scatter and slope is explained later (pages 75–78). For an interesting historical account of the development of regression and correlation concepts, see J. P. Guilford and B. Fruchter, *Fundamental Statistics in Psychology and Education*, 6th ed. (New York: McGraw-Hill, 1977).

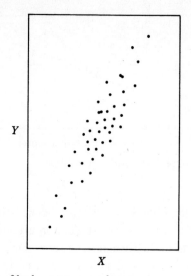

Y

X

Figure 6-4   Scatterplot of high positive correlation.

slope, the greater the scatter, and the smaller the scatter, the greater the slope. When scores are *not* in standard units, the slope is determined mostly by the whim of the graph maker. We shall return to that issue on pages 75–78, as we complete our discussion of graphic representations.

If, as in real life, a correlation is not perfect, there is always some scatter. But if the X-Y relationship is extremely strong, most of the deviations from the regression line are small (see Figure 6-4); we can make very accurate predictions of Y scores from a knowledge of X scores. At the other extreme is the random X-Y relationship depicted in Figure 6-5, where there is no tendency for a low Y to be associated with a low X or a high Y with a high X, and as you can see, there is a near maximum of scatter. Knowing a person's score on X does not help at all in estimating that person's score on Y.

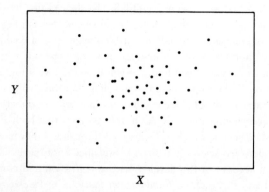

Y

X

Figure 6-5   Scatterplot of zero correlation.

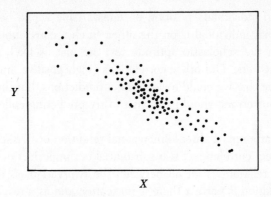

Figure 6-6 Scatterplot of high negative correlation.

Negative correlations behave in exactly the same way as the positive ones except that in each instance the regression line slopes *down* from left to right instead of up. Figures 6-6 and 6-7 are examples.

Figures 6-3 through 6-7 have been contrived to represent perfect positive, high positive, zero, high negative, and perfect negative correlations, respectively. Either of the perfect correlations would enable us to predict *Y* precisely from a knowledge of *X*, and vice versa. The zero correlation would tell us that a

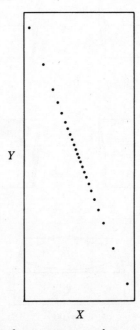

Figure 6-7 Scatterplot of perfect negative correlation.

knowledge of an individual's position on one variable would be useless in predicting where that individual is on the other. Such a correlation would say that
high scorers on the scholastic aptitude test are just as likely to flunk out of
school as low scorers. The other correlations (high positive and high negative)
would not ensure that we could make perfect predictions, but they would significantly reduce our errors, should we wish to try predicting college performance
from test scores.

Figure 6-8 attempts a three-dimensional rendition of five scatterplots. It includes two perfect correlations, two substantial but imperfect ones, and one plot
that reveals no relationship at all between the two variables. You can see that
when the correlation is perfect there is *no* scatter, and as a result the scores are

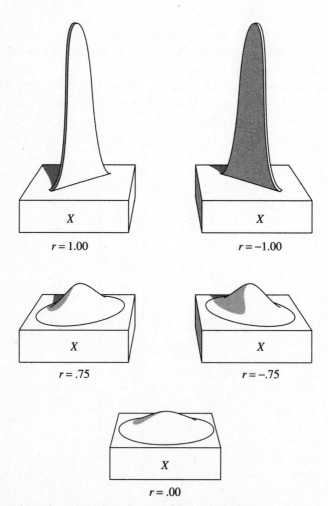

Figure 6-8   Three-dimensional renderings of five scatterplots.

piled high. In stark contrast, a zero correlation spreads scores out over most of the horizontal surface, with relatively few in any one place. Other correlations approximate those two in varying degrees. The relationship represented by $r =$ .75, for example, is far from perfect; but it is also a long way from zero. The relationship that it reveals may be interesting for itself, and it also may have practical consequences. The prime illustration of the latter is the fact that with a correlation that high and a knowledge of X you could substantially reduce your errors in predicting Y.

A distribution need not be normal to support calculating an r. It need only be unimodal and fairly symmetrical. And the regression line must be approximately straight. Always construct a scattergram, whether or not you calculate a coefficient of correlation. The Pearson r, based as it is on standard scores, guards against distortions of scale, which the scatterplot does not; but the scatterplot may detect configurational anomalies that would be missed without it—asymmetry, for example, a curved regression line, or even more than one line. The coefficient will not catch any of these.

Effect of Restricted Variability

Everything I have said about correlation is based upon the assumption that the individuals measured constitute random samples from both distributions. If the correlation is between IQ and scholastic achievement, for example, we should select a large sample of subjects and measure each of them twice, once for IQ and once for achievement. Their scores on the IQ test should form a distribution similar in every way but size (n) to the one that would have emerged had we tested an entire population (N) in the same way, and similarly for achievement.

On the other hand, it is sometimes useful to subdivide a general population into subcategories, and it is legitimate to call each subcategory a population, too. For example, instead of including literally everybody in a population of IQs, we can arbitrarily define a population that includes only college students. Now we will study *that* population (or take our sample from it).

All of this is fine as far as it goes, but we must recognize that subdividing a population can have a significant effect on the correlation coefficient. The best way to demonstrate that fact is again to draw a diagram; Figure 6-9 is appropriate to our present need. It shows a hypothetical relationship between intelligence (as measured by a standardized test) and achievement (as measured by grade point average) in the general population. It also shows what the relationship would be if the population were defined as "applicants to graduate schools." The entire range of IQ scores for the general population forms the baseline of the diagram, and the entire range of grade point averages forms the ordinate. Brackets I and II enclose those segments on each scale within which 90 percent of applicants to graduate school may be found.

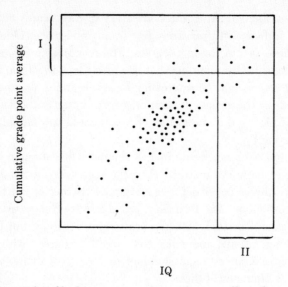

Figure 6-9   Scatterplot of high positive correlation showing effect of restricted variability.

The scatterplot in Figure 6-9 reveals a substantial correlation between intelligence test scores and scholastic achievement. However—and this is the point toward which this discussion has been directed—a plot of the area marked off by segments I and II reveals no relationship at all. Figure 6-10 is an enlargement of that area.

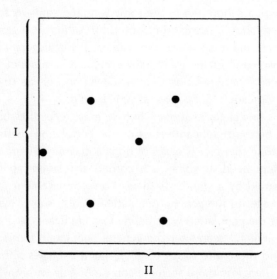

Figure 6-10   Restricted area (enlarged) of scatterplot in Figure 6-9.

Be cautious, then, whenever you interpret a low correlation coefficient. In the above illustration, it would be proper to say that among applicants to graduate schools, there is virtually no relationship between IQ and grade point average, but be careful *not* to infer from that *a similar lack of relationship in the general population.* A correlation coefficient is relevant only to the population that you study directly or from which your sample is drawn.

Standard Scores in Correlation

While introducing the concept of standard deviation back in Chapter 4, I mentioned that among the common statistics to which it lends itself is the product-moment coefficient of correlation. This section will show you how the standard deviation functions in the process of arriving at a correlation coefficient, and at the same time it will deepen your comprehension of the *concept* of correlation.

The defining equation for the coefficient [Formula (6-2), page 66] is repeated here for your convenience:

$$r = \frac{\Sigma xy}{nS_x S_y}$$

or, to help clarify my next point,

$$r = \frac{\Sigma x \; y}{n \; S_x \; S_y}$$

Now according to Formula (5-1), a standard score (z) is a deviation score divided by the standard deviation of the distribution in which it is found; in other words, it is a deviation score in standard deviation units. In Formula (6-2), we have two such scores:

$$\frac{x}{S_x} \text{ and } \frac{y}{S_y} = z_x \text{ and } z_y$$

so we can rewrite the formula in terms of standard (z) scores:

$$r = \frac{\Sigma z_x z_y}{n} \tag{6-3}$$

If you remember that Σ/n is how you compute any mean, it should now be clear to you that r is the mean of the cross products of x and y when *both are expressed in standard units* (z_x and z_y). It is also *the slope of the regression of the Y variable on the X*, again when both are in standard units.

If we make scatterplots of 10 raw-score relationships that are all of the same strength and direction, we may get 10 different slopes of our regression lines; if we first convert all measurements to standard scores, we get identical slopes. The slopes in Figures 6-3 and 6-7 are positive and negative, respectively, but they are both perfect relationships; if they were plotted in standard units, the degree (though of course not the direction) of their regression slopes would be the same. Without the conversion, they are very different.

Figure 6-11 depicts those relationships in a slightly different way. Of the four lines that cross diagrams A, B, and C, one line (the perpendicular) is at the mean of all the X scores and another line (the horizontal) is the mean of the Ys. The third and fourth lines (the bold diagonals) are regression lines.

The reason that you don't see four lines at $r = 1.00$ (diagram A) and $r = .00$ (diagram C) is that when one line corresponds exactly to another, you can see only one. In diagram A, where the correlation is perfect, the regressions of Y on X and of X on Y are merged into one line that is just 45° from each of the two baselines. In diagram C, where $r = .00$, the regression lines are separate, but the regression of Y on X is at the mean of Y all the way across, and that of X on Y is at the mean of X; so the two regression lines are superimposed on the two mean lines. (They *are* the two mean lines.) Thus, you see two lines instead of four.*

Have you noticed that in contrast to Figures 6-3 through 6-7, pages 69–71, all the scatterplot grids in Figure 6-11 are precisely square? There is a very good reason for that: The X and Y variables are plotted in the same units, namely, standard deviations, and a rectangle with equal sides is a square.

There is one feature of Figure 6-11 that could be misleading if it were not explained. That feature is especially noticeable in diagram A, in which it appears that there is, after all, some scatter when the correlation is perfect ($r = +1.00$).

That is an erroneous impression. What has happened is that dots that should have been stacked directly on top of one another have been dispersed on the plane of the page so that they can all be seen; when you are viewing a stack of identical objects from a point directly above them, all you can see is the one that sits at the top of the stack. Figure 6-8, page 72, is an attempt to circumvent that difficulty by rendering the plots in three dimensions. It might be helpful to look at those drawings again now.

You can see there that when the correlation is perfect there is *no* scatter, and the scores are piled high as a result. In stark contrast, a zero correlation spreads out over most of the board, with few scores in any one place. Other correlations approximate those two in varying degrees.

---

*If you will turn Figure 6-11 over on its side and examine the regression of X on Y, you will discover that in every diagram it has exactly the same slope as that of Y on X. (The direction of the slope is reversed because the sequence of Y scores is reversed when the diagram is viewed from the side.)

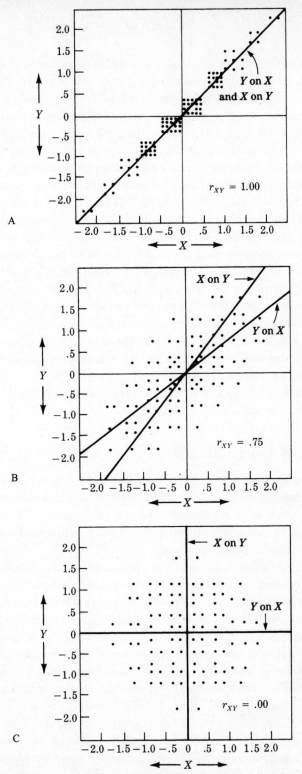

Figure 6-11 Three scatterplots on standard grids (all scores in standard units): (A) one regression line, (B) two regression lines, and (C) two regression lines. See also Figure 6-8.

If your plot is done with raw scores, the length of each unit on either dimension is arbitrary. For example, you may use inches or feet on one dimension and ounces or pounds on the other, and you may let 1 inch, foot, ounce, or pound be represented by whatever distance is convenient, depending on the kind of graph paper you are using and the horizontal and vertical space that is available. Furthermore, if you wanted to deceive someone by manipulating the slope of a regression line, you could do it very easily by simply selecting scale units that would suit your purpose. But standard scores keep you honest; once the conversion to standard scores has been made, for a given strength of relationship, there is one and only one degree of slope.

When no relationship exists, the slope is zero; since there is no reason to expect the Y scores accompanying any one X value to be any different from those associated with any other, the best estimated central tendency of the Ys at any given value of X is simply the mean of all the Ys in the entire distribution. So as our regression line moves from left to right on the X-axis, it moves not at all on the Y; which is to say, the slope of the regression line is zero, as in Figure 6-11C. For a perfect relationship, it is 1.00, as in Figure 6-11A. All other relationships are somewhere between these two extremes. (Remember that a correlation of -1.00 is as perfect as a correlation of 1.00.)

In short, one meaningful interpretation of r is as *the slope of the regression of Y on X when both are laid out in standard units.*

A Matrix of Correlations

Investigators seldom publish studies that report but a single coefficient of correlation. Many published studies cite tens and some even hundreds of such coefficients. There needs to be some systematic way of recording all that information so that any part of it can be seen and compared to other parts. The system commonly used is called a *correlation matrix.*

A matrix is a table consisting of rows and columns of numbers. A matrix of correlation coefficients differs from most others in that it is perfectly symmetrical across its diagonal; that is, the numbers in the upper right half form a mirror image of those in the lower left. A single example should suffice to clarify the concept.

The ultimate objective of many psychological studies is to reveal the structure of human behavior and experience. One approach to that objective is correlational. Investigators take measurements and correlate the results, thus discovering what is related to what. Frequently the number of variables in the resulting table (the matrix) is enormous, but the *form* of the table can be illustrated just as well with only a few. Table 6-5 is a complete matrix of the coefficients that might result from the correlation of five variables, each one with every other one.

The coefficient of correlation between any two variables appears at the intersection of a row and a column, the row representing one of the two variables,

TABLE 6-5 Intercorrelations of five ability measures: the complete matrix

		Shop work quality	Mechanical assembly	Mechanical information	IQ	School grades
		1	2	3	4	5
Shop work quality	1	1.00	.60	.40	.20	.40
Mechanical assembly	2	.60	1.00	.40	.00	.10
Mechanical information	3	.40	.40	1.00	.70	.50
IQ	4	.20	.00	.70	1.00	.60
School grades	5	.40	.10	.50	.60	1.00

Source: Adapted from P. E. Vernon. *The Structure of Human Abilities.* London: Methuen & Company, Ltd, 1950, p. 102.

the column representing the other. A brief scrutiny of the matrix will reveal how that comes about and how the table provides a systematic way of recording the results of a correlational study. You may also notice the special feature of matrices that I mentioned earlier: the perfect symmetry across the diagonal.

The fact is, however, that you don't need the symmetry, and you don't need the diagonal. The latter is derived not from data but from theory: Each *1.00* is an idealized reliability coefficient—that is, it shows how each test would correlate with itself if it were perfectly reliable. That information is useful only as a reference point—or rather, a reference line—because you know in advance that a perfectly reliable test would correlate 1.00 with itself. (See the section after this one.) The symmetry is interesting, but it results from the incorporation of redundant information: If you know what is on one side of the diagonal, you know what is on the other.

Table 6-6 omits everything (1) that you know in advance and (2) that you can infer with certainty by examining the entries on the other side of the diagonal. That is the form in which reports are commonly made (although the diagonal series of perfect correlations is frequently not deleted). The essential information is all there, namely, the intercorrelations among five measures taken of a large number of workers in automobile repair shops. Scores on *shop work quality* are derived from supervisors' ratings; *mechanical assembly* is a test that confronts examinees with a dismembered machine and requires them to reassemble it. The

TABLE 6-6 Intercorrelations of five ability measures: an abbreviated report

		1	2	3	4	5
Shop work quality	1					
Mechanical assembly	2	.60				
Mechanical information	3	.40	.40			
IQ	4	.20	.00	.70		
School grades	5	.40	.10	.50	.60	

Source: Adapted from P. E. Vernon. *The Structure of Human Abilities*. London: Methuen & Company, Ltd, 1950, p. 102.

data for the *mechanical information* entries are also test scores, as are those for *IQ*. *School grades* are overall averages (means) from high school.

Now look again at Table 6-6, this time for content. Think about it. Think about the relative magnitudes of the relations represented by these coefficients. (All the relations are positive, which is nearly always the case with abilities; so you needn't be concerned with *direction* when making these comparisons.) Are your preconceptions generally confirmed? Are there any surprises? Do any possible explanations occur to you concerning counterintuitive results? These are the kinds of thoughts that form in the mind of a good researcher in the presence of a correlation matrix.

Expectancy Tables and Predictive Validity

There are many occasions in contemporary medical and social science research on which a relation between *two* variables is the focus of interest. On such occasions, various correlation techniques may be applied. When a correlation coefficient is used to predict scores on one variable from scores on another, the accuracy of the prediction is known as *predictive validity.*

The Expectancy Table as a Scatterplot

A correlation coefficient can be a very useful statistic when the relation between two variables is of interest. However, if you need to communicate relational information to a person untrained in statistics (for instance, a high school student or the student's parent) some other way must be found—some display that is concrete enough to be comprehended without benefit of previous study. A scatterplot is one such display. You will recall that in seeking the Pearson *r* we were really attempting to define the slope of the regression line with both X and Y scores in standard units. The formula for r converts the scores into standard units and makes possible a very concise communication (r) to *anyone who is familiar with the process.* But by using a lot more space, we can communicate much of the same information without converting to standard scores. A scatterplot displays that information in a way that makes it accessible to almost anyone.

Figure 6-12 shows the regression of grade point averages on test scores. The line across the drawing is the regression line—that is, the *line of best fit.* (In this case, the slope is .37, as it would have been if the two scales had been laid out in standard units.) It would be easy to predict grade point averages from test scores by simply referring to that line. The only trouble with such predictions is that they leave out some important information: They ignore *variability.* If X were an aptitude test score and Y were grade point average at Central University, the regression line would tell a student with a score of 57, for example, that in college his grade point average would be 2.2. The entire scatterplot, on the other hand, would say that although a 2.2 grade point average might be the central tendency of students who scored 57 on the test, it would most assuredly not be the *only* possibility.

An *expectancy table* functions in the same way as a scatterplot but is easier to use. The space in the table is divided on two dimensions into rows and columns, thus forming an array of cells. If the entry in each cell is a frequency, then the scatterplot is almost exactly duplicated, as in Table 6-7. Usually, however, the frequencies in each column are transformed into percentages of the column sum, as

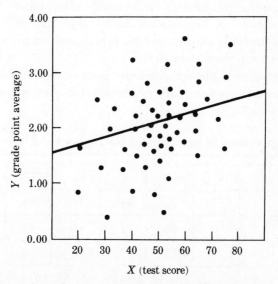

Figure 6-12   Scatterplot with superimposed regression line when $r = .37$.

TABLE 6-7 Expectancy of frequencies for Figure 6-12 ($r = .37$)

Y	10–19	20–29	30–39	40–49	50–59	60–69	70–79	80–89
3.00–3.99				1	2	1	1	
2.00–2.99		1	2	7	7	3	2	
1.00–1.99		2	3	7	7	2	1	
0.00–0.99		1	1	2	1			

in Table 6-8. In either case, you enter the table with the raw score from a student's test and emerge with more than one possible grade point average.

Using an Expectancy Table

Now you can tell your student not only his most probable grade point average in college, but also his chances of doing better (or worse) than the most probable performance. Let's say that Joe Cawledge's score on test X is 57; Table 6-8 says that of all students who previously scored in the 50s and subsequently attended Central University, 6 percent earned a grade point average below 1.00, 41 percent a GPA in the interval 1.00 through 1.99, another 41 percent between 2.00 and 2.99, and 12 percent a 3.00 or better. That kind of information would be comprehensible to any interested person. Indeed, it may be helpful even to the experienced counselor, for although the counselor knows immediately that any r below a magnitude of 1.00 implies some scatter, it is often difficult to visualize the *amount* of scatter. The expectancy table gets right down to cases (as does, in a slightly different way, the scatterplot).

A refinement often made in the interpretation of such a table is the derivation of *cumulative* percentages. Table 6-9, unlike Table 6.8, implies not that Joe

TABLE 6-8 Expectancy of percentages for Figure 6-12 ($r = .37$)

Y	10–19	20–29	30–39	40–49	50–59	60–69	70–79	80–89
3.00–3.99				6	12	17	25	
2.00–2.99		25	33	41	41	50	50	
1.00–1.99		50	50	41	41	33	25	
0.00–0.99		25	17	12	6	0	0	

TABLE 6-9   Expectancy of cumulative percentages for Figure 6-12 ($r = .37$)

	10–19	20–29	30–39	40–49	50–59	60–69	70–79	80–89
≥ 3.00				6	12	17	25	
≥ 2.00		25	33	47	53	67	75	
≥ 1.00		75	83	88	94	100	100	
≥ 0.00		100	100	100	100	100	100	

Y (rows); X (columns)

Cawledge has a 41 percent chance of achieving a grade point average between 2.00 and 2.99, but that the probability of his doing *2.00 or better* is .41 + .12 = .53. That sum is often more useful than either probability by itself—so often, in fact, that many expectancy tables are made up entirely of cumulative percentages. Table 6-9 is based upon the same data as Table 6-8; the only difference is that all the percentages are cumulative. Although that structure makes Table 6-9 less flexible than Table 6-8, the many persons who are interested primarily in cumulative percentages will find it easier to use.

Reliability and Validity

At the beginning of this chapter, I suggested that an index of correlation would allow us to check on the accuracy of predictions made by a scholastic aptitude test. Now, by substituting the test score for X and college grade point average for Y in each of the diagrams in Figure 6-11, you can see just how accurate the predictions would be if the correlation were .00, .75, or 1.00.

A correlation coefficient that is used to predict a criterion score (Y) from a test score (X), as in the case of the scholastic aptitude test, is called a *validity coefficient*.* The *criterion* is an index of the entity that our test is supposed to be measuring—in this case, scholastic aptitude, or the capacity to profit from schooling. (Grade point average is generally accepted, though sometimes reluctantly, as an index of scholastic achievement—that is, the extent to which an individual has profited from schooling.)

If we were to administer our scholastic aptitude test to all entering freshmen as indicated earlier, and if we were to test them again a week later, we could compute an r that would tell us the relationship between the first and second

---

*The examples given are of predictive validity and test–retest reliability. There are other kinds of validity and of reliability, but these examples illustrate how correlation coefficients can represent the relationships involved.

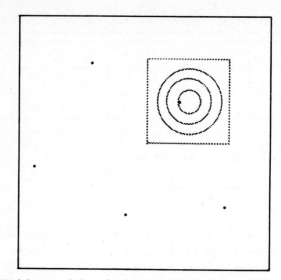

Figure 6-13 Widely spaced shots fired at invisible target concealed by sheet of white gauze.

testings. Instead of an r_{xy}, we would have an r_{xx}, and it would be called a *reliability* coefficient because it would tell us the extent to which we can depend on the test to give us the same results from one administration to the next. (You can't measure anything reliably with a rubber yardstick!)

The rubber yardstick metaphor is a good one to help you remember the concept of reliability, but it bears no direct relation to *validity*. Here is an analogy that illustrates both reliability and validity: Imagine a marksman aiming a handgun in the general direction of a target that is a hundred feet away. I say "in the general direction" because there is a huge sheet of white gauze that hides not only the target but a considerable area around it. Our marksman takes aim at a spot that he hopes is the bull's eye. He fires five shots—and scatters them all over that cloth screen. Figure 6-13 shows the pattern that the shots make on the screen and on the hidden target.

Another marksman takes aim at the same spot as the first one did. She groups her shots as shown in Figure 6-14, thereby missing the target completely.

A third marksman knows where the target is, takes aim in that direction, and pumps off five shots that are as close together as those of the second marksman. The result is shown in Figure 6-15.

Now review the concepts of reliability and validity, and see if you can tell which of them is illustrated by which marksman. Reliability? That would have to be the closely grouped patterns, wouldn't it? Shot after shot lands in nearly the same place; you can count on it, just as you can count on a reliable test to yield nearly the same scores when the same people are retested. Proximity to the target, on the other hand, is the analog of validity. Close grouping of shots doesn't

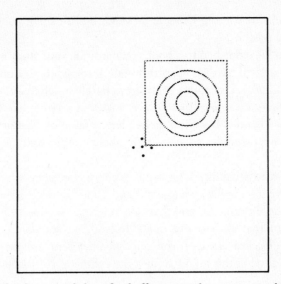

Figure 6-14   Closely grouped shots fired off-target with target concealed by sheet of white gauze.

help a bit unless it is on target. Reliable testing is useless unless we know what we are testing.

So high reliability does not ensure high validity. But it is also important to note that *low* reliability absolutely precludes it. Unless your shots are all very close together, it is impossible to score well. If a test has low reliability, it can have *some* validity (see Figure 6-13) but not much. To put it another way, a test

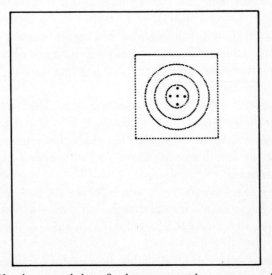

Figure 6-15   Closely grouped shots fired on-target with target concealed by sheet of white gauze.

can be highly reliable without being at all valid, but it cannot be very valid without being highly reliable.

If when you have finished this book you continue your study of statistics and measurement, you will have an opportunity to develop a better understanding of the kinds of validity and reliability that I have barely touched upon; you will also discover that there are others that I have not even mentioned. We will not go further into those matters here, but do be alert to later applications. Even in this book, there are opportunities to learn more about validity and much more about reliability.

With respect to reliability, Chapter 8 is about precision, which is really another way of saying reliability. Chapters 9 and 10 are concerned with the significance of obtained differences, and that, too, answers the question "How reliable is the information that we have obtained?" Indeed, any standard error (as in *standard error of the mean* or *standard error of a difference between means*, to name the two specifically cited in this book) implies the estimation of reliability.

Summary

It is often important to know the relationship between two variables. *Coefficients of correlation* are indices of relationship. Many such indices have been devised for various applications, but we have discussed just two: the rank-difference and the product-moment coefficients.

Any correlation coefficient is an index of the extent to which measurements of the same individuals are to be found on corresponding segments (e.g., low, middle, and high) of two different scales. When such a relationship holds precisely and with no exceptions, the correlation is said to be perfect, and the coefficient is 1.00. Less-than-perfect relationships produce coefficients smaller than 1.00. A *negative* correlation is just as strong as a *positive* correlation if its coefficient is equally large; the coefficient indexes magnitude and direction separately in a single number that varies from .00 to 1.00 and that does or does not carry a minus sign.

The Spearman *rank-difference* (ρ) procedure places all subjects in order from smallest to largest on the basis of their scores on X; then it ascertains how nearly their Y scores approximate that order. The better the approximation, the smaller are the differences in rank (hence the name "rank difference") and the larger the coefficient of correlation.

The Pearson *product-moment coefficient* (r) is like the rank-difference coefficient except that it retains information that the latter throws away—namely, the distances that separate subjects of adjacent ranks on either of the variables being correlated.

If the position of each subject (specified by a pair of observations) is plotted on both X and Y coordinates, a graph called a *scatterplot* emerges. In that graph it is possible to perceive directly the degree and direction of the relationship between two variables. (We might still want to compute an *r*, however; the most *direct* index of an entity is often not the most *precise*.) The same general pattern, expressed somewhat more crudely in tabular form, has been named an *expectancy table* after its intended use in the prediction of outcomes.

The formula for *r* features in its numerator the deviation scores of both *X* and *Y* variables, but the presence of their standard deviations in the denominator is important, too, for they have the effect of converting deviation scores to standard scores. The Pearson *r* is the mean of the cross products of deviation scores on both variables when those deviation scores are expressed in standard units. The conversion to standard units also permits another definition of *r*: It is the slope of the regression of the *Y* variable on the *X* when both have been converted to standard scores.

Correlational studies seldom stop at two variables. Some relate tens and even hundreds of them, each one to each of the others. Whenever more than one pair of variables is involved, a convenient way of displaying the coefficients is the *correlation matrix,* in which all the variables are listed both horizontally, labeling the columns of cells in the table (matrix), and vertically, labeling the rows. It is a good way to get an overview of all the relationships being studied.

Reliability is one of a pair of important attributes that characterize all measurements; the other is *validity.*

In the kinds of situations cited here, the *coefficient of reliability* tells us whether a test does whatever it does *consistently;* the *coefficient of validity* tells us whether it is doing what we have *believed* it to be doing.

A coefficient of correlation quantifies the empirical relation between two variables. But does an *empirical* relation imply a *causal* relation? Chapter 11 will consider that question by comparing correlational with experimental studies.

Sample Applications

EDUCATION

You are an elementary school consultant helping teachers plan activities to enhance students' emotional and social development. You have observed informally that students who have a negative view of themselves (low self-concept) seem not to become involved in helping class groups achieve common goals (low social responsibility). You wonder whether those observations can be confirmed objectively. To investigate this relationship you administer a self-concept and a social-responsibility scale to 200 fourth-, fifth-, and sixth-grade students. After the data have been collected, how do you organize them to answer your question about the relationship?

POLITICAL SCIENCE

Researchers have frequently asked whether there is any relationship between the amount of domestic conflict within a given country (X) and the amount of foreign conflict which that country initiates (Y). Assume that you have constructed a conflict scale and collected data for 50 countries on the values of both X and Y. What statistic will answer your question?

PSYCHOLOGY

You are a child psychologist interested in the possible positive relationship between the amount of sugar in the breakfast diet of young children and hyperactivity (i.e., inattentiveness, excessive muscular activity, etc.). You ask 100 mothers of elementary school children to keep records on what their children eat and drink each morning for breakfast. A nutritionist then analyzes the parental reports and notes the average weekly amount of sugar ingested. Data on hyperactivity are gathered by behavior analysts who visit classrooms on a daily basis and rate each child on a 10-point continuum. What statistic will indicate the strength and direction of the relationship between sugar intake and activity level?

SOCIAL WORK

You are a social worker studying the records of women who receive financial assistance from your state's welfare agency. You are interested in a possible association between a woman's level of self-esteem and the number of months she receives benefits. What single index can summarize information pertinent to this problem?

SOCIOLOGY

Your research class is locked in a disagreement over the possible relationship between the conservatism/liberalism of academic disciplines and authoritarianism. (A study of authoritarianism has already been done; see the sociology application in Chapter 5.) The question is whether liberal arts disciplines such as English, history, philosophy, and psychology either attract or produce professors who are significantly more liberal and less authoritarian than professors of disciplines such as biology, business, chemistry, math, and physics, who, some students claim, are more conservative and authoritarian.

Having chosen this issue as your research topic, you begin your project by acquiring the data from the authoritarianism study. Then, to identify the conservative/liberal orientation of academic disciplines, you pick a panel of 10 judges to place the disciplines (not the professors) on an interval scale from 0 (most conservative) to 100 (most liberal).

How do you quantify the relation between these two variables?

7

Description to Inference: A Transition

In Chapter 1, I pointed out that statistics has two main functions: (1) *description* of populations, or of samples taken from those populations, and (2) *inference* from the properties of samples (*statistics*) to those of populations (*parameters*). If you are doing research for your employers and for their eyes only, description may be all you will need. But if you intend to generalize the outcome beyond your own situation, you will need something more, for in that event the subjects of your study are but a sample of a larger population. It is then that you will need inferential, sometimes called "sampling," statistics.

Chapter 1 identified description and inference as the two main functions of statistics. Since then you have seen references to both samples and populations in the chapters on *central tendency* and *variability*. In this chapter we shall address those topics once again, this time at a higher level of discourse.

In each of the first two first-order headings of this short chapter, most of the terms are symbols rather than verbal descriptions. For the first title I could have written, "Describing Observed Distributions via the Mean of the Sample and via the Mean of the Population, and Estimating the Latter from the Former"; and for the second, "Describing Observed Distributions via the Standard Deviation of the Sample and via the Standard Deviation of the Population, and Estimating the Latter from the Former." I wanted each title to serve as a kind of outline of what follows it. Both versions do that (and both are rather intimidating, I'm afraid), but the words-only version would have consumed six lines instead of two and would not have been much easier to interpret. Actually, the briefer form makes a better summary. Practice interpreting each title again after you have studied the text, and I think you will agree.

Before proceeding further, and again when you have finished reading both sections, compare the two titles as they are written and note their structure:

Their contents differ, but their form is exactly the same. That may help you to comprehend and remember both the form and the content.

Describing Observed Distributions via $\overline{X}$ and μ, and Estimating μ from $\overline{X}$

We begin with a discussion of description and of inference (estimation) with respect to a measure of central tendency—specifically, the mean.

Describing Observed Distributions via $\overline{X}$ and μ

This section will be short, mainly because Chapter 3 showed you that the formula for the population mean (μ) is identical to that of a sample mean ($\overline{X}$). To put it another way,

$$\overline{X} = \frac{\Sigma X}{n}$$

where ΣX is the sum of the (raw) scores in a sample and n is the number of such scores. Similarly,

$$\mu = \frac{\Sigma X}{N}$$

where ΣX is the sum of scores in the *population* and N is the number of such scores. If after calculating a sample mean you were suddenly informed that your "sample" is really the *population* of interest, no recalculation would be necessary; you would simply report your mean as μ instead of $\overline{X}$.

Estimating μ from $\overline{X}$

There is a much more important implication of learning that your "sample" is really the whole population: You need not be concerned about the hazards of making inferences about a population you have not been privileged to observe.

If, on the other hand, your "sample" really *is* but a sample of the target population, you *do* have to be concerned: You need (1) to select the best available estimates of important parameters and (2) to ascertain the *reliability* of such estimates.

With respect to the parameter known as the mean, your first concern—the selection of the best estimate—is easily abated: *The best estimate of the population mean is the mean of a sample* taken from that population.

With respect to ascertaining the reliability of your estimate, the answer is not so simple. It involves a new concept: *the standard error of the mean,* which is a measure of the variability of a distribution of sample means—a notion that I shall develop for you in Chapter 8. The variability of sample means is significantly dependent on that of individual scores in the population, however; so we must deal with that first.

Describing Observed Distributions via S and σ, and Estimating σ from S

In the preceding section I asserted that the best estimate of the population mean is the mean of the sample. The state of affairs with respect to the standard deviation, however, is not nearly so simple.

Describing Observed Distributions: S and σ

You may recall that as a kind of recapitulation of our discussion of the standard deviation in Chapter 4, I listed four formulas that are related in one way or another to the concept of the standard deviation. I have reproduced them here for your convenience:

$$\text{The mean of the raw scores} = \frac{\Sigma X}{n}$$

$$\text{The average deviation} = \frac{\Sigma |x|}{n}$$

$$\text{The variance} = \frac{\Sigma x^2}{n}$$

$$\text{The standard deviation} = \sqrt{\frac{\Sigma x^2}{n}}$$

The numbers that enter these formulas are taken directly from your sample, and the numbers that emerge from them refer directly to that sample. The very word "sample" implies, however, that your interest is not primarily in the sample. Rather, it is in the *population* from which the sample was drawn: It is in *parameters* rather than *statistics* per se. Indeed, when the population is small enough to be measured in its entirety, you won't even *draw* a sample.

In Chapter 1 we compared description to inference, and I suggested several examples of each. Some of those examples appear in Table 7-1, along with comments concerning the description or inference involved. Study that table.

One implication of this discussion is that the standard deviation of a sample (S) is of only theoretical importance, because whenever you draw a sample from a population, your primary interest is always in the population, not the sample. Except for its theoretical function, a sample is useful only insofar as it helps you

TABLE 7-1   Some examples of description and inference[a]

	Example	Comment
1	A college professor tests a class for knowledge of subject matter.	Description of *entire* population; namely, that particular class. Inference not necessary.
2	A political polling agency interviews 1000 potential voters in a national election.	Inference (of preferences) to *entire* electorate from those of sample.
3	A national election is held.	Description of population, namely, the *entire* electorate.
4	A pharmacologist supervises clinical trials of a new drug on 500 patients.	Inference (of a specified effect of the drug) to *all* potential patients. Description of population impossible.
5	The current batting average of every player is made available to the broadcasters of a baseball game.	Description of *all* at-bats. Description possible here, even though there may be a very large number of at-bats, because everyone has been observed and the result recorded.

[a]Samples are always described, but the properties of a population may be inferred from those of a sample.

to learn something about the population. Unfortunately, a sample's standard deviation (S) is a biased estimate of the standard deviation of its parent population (σ).* So now that you thoroughly comprehend S, I am telling you that it should never be used!

"But," you should be asking about now, "what about situations in which I can observe the *entire population,* as in examples 1, 3, and 5 in Table 7-1? What formula will give me the population standard deviation in those situations?"

The answer is filed somewhere in your memory. If you cannot recall it immediately, turn back to Formulas (4-2) and (4-3), pages 38–39. A quick review will remind you that *the two formulas are identical* (excepting for the symbols that indicate references to sample and population, respectively). That makes sense, because in both cases you are describing a set of actual observations, as distinguished from inferring the nature of a set of observations that have not been made.

*If we were to calculate an S for each of an infinite number of samples, the mean of those Ss would not be equal to σ.

Estimating σ from S: A New Statistic, s

When you do want to estimate some characteristic (parameter) of a population that, except for a small sample, you have *not* observed, the situation is different. As I mentioned earlier, the standard deviation of a sample (S) is a biased estimate of the standard deviation of the population (σ). Specifically, S tends to underestimate σ.

The sample standard deviation is the closest approximation of the population standard deviation, however, so we'll start with that and then modify it to correct its tendency to underestimate. Diminishing the denominator (n) of the fraction within the formula

$$S = \sqrt{\frac{\Sigma x^2_{sample}}{n}}$$

would produce a value larger than S, and that is what we do: The formula for the estimated standard deviation of a population is

$$s = \sqrt{\frac{\Sigma x^2_{sample}}{n-1}} \qquad\qquad (7\text{-}1)$$

where s is the standard deviation of the population as estimated from sample data, Σx^2_{sample} is the sum of the squared deviations in the sample, and $n - 1$ is *degrees of freedom*.

Degrees of Freedom

The term *degrees of freedom* refers to the number of scores that are free to vary. The examples below should clarify the phrase "free to vary." Imagine a very simple situation in which the individual scores that make up a distribution are 3, 4, 5, 6, and 7. If you are asked to tell what the first score is without having seen it, the best you can do is a wild guess, because the first score could be *any* number. If you are told the first score (3) and then asked to give the second, you probably won't guess some number in the millions, but logically there is no reason why it should *not* be in the millions or even higher; it, too, can be any number. The same is true of the third and fourth scores; each of them has complete "freedom" to vary. But if you know those first four scores (3, 4, 5, and 6) *and you know the mean* of the distribution (5), then the last score can only be 7. If, instead of the mean and the 3, 4, 5, and 6, you were given the mean and, say, 3, 5, 6, and 7, the missing score could only be 4. A few examples of your own will help you to grasp this idea. In every case, if the mean is known, the missing score is *determined* by your knowledge of the other four, and the degrees of freedom of all the scores taken together has been reduced by 1; hence the term $n - 1$ in the formula for the estimated standard deviation.[1]

Degrees of freedom in any calculation is the number of values that are free to vary, given whatever mathematical restrictions are inherent in that calculation. The estimation of a population standard deviation involves one such restriction—pre-knowledge of the mean—and the consequent loss of 1 degree of freedom. However, that is not the only possibility; *df* can be *n* (where there are *no* restrictions), $n - 1$ (as in the present case), $n - 2$, $n - 3$, or even smaller.

To summarize: When a formula is concerned with straightforward *description*, degrees of freedom is *n*. When the objective of a calculation is *inference*, some restrictions apply, so that degrees of freedom is smaller than *n* (e.g., $n - 1$, $n - 2$, $n - 3$, etc.). In every case the amount subtracted from the denominator of a *statistic's* formula is large enough to compensate for that formula's tendency to underestimate the corresponding *parameter*.

You may expect to see more instances of *df* smaller than *n*, because from here on there will be much ado about estimation (inference).

SUMMARY

Earlier chapters demonstrated sophisticated ways to *describe* a distribution of individual measurements; this one prepares you for the later chapters where such data are used to *infer* properties of a distribution not in evidence. The preparation is done in terms of two kinds of measure—a measure of central tendency (the mean) and a measure of variability (the standard deviation).

The message with respect to the mean is simple: The mean of any sample from the population of interest is an unbiased estimate of the population mean.

The case of the standard deviation is more complicated, but it can be abridged as follows:

1. The standard deviation of a sample

$$S = \sqrt{\frac{\Sigma x^2_{sample}}{n}}$$

    can be calculated directly from the data; but you will have no occasion to do so, because when you take a sample of a population, your interest is in the population.

2. The standard deviation of the population

$$\sigma = \sqrt{\frac{\Sigma x^2_{pop.}}{N}}$$

    is calculated directly from data whenever it is feasible to do so. Its formula is essentially the same as that of S.

3. S underestimates σ.

TABLE 7-2   Six important concepts with their symbols and formulas[a]

$\overline{X}$ = mean of a sample	$\dfrac{\Sigma X_{sample}}{n}$
μ = mean of the population	$\dfrac{\Sigma X_{pop}}{N}$
$\overline{X}$ = mean of the population as estimated from a sample	$\dfrac{\Sigma X_{sample}}{n}$
S = standard deviation of a sample	$\sqrt{\dfrac{\Sigma x_{sample}^2}{n}}$
σ = standard deviation of the population	$\sqrt{\dfrac{\Sigma x_{pop.}^2}{N}}$
s = standard deviation of the population as estimated from a sample	$\sqrt{\dfrac{\Sigma x_{sample}^2}{n-1}}$

[a]If the word "estimated" does not appear opposite a symbol, the value it represents is calculated directly from data.

4. The standard deviation of the population as estimated from that of a sample compensates for that underestimation. In other words,

$$s = \sqrt{\frac{\Sigma x_{sample}^2}{n-1}}$$

is the same as S except for the loss of 1 degree of freedom (the "-1" in the denominator). Decreasing the denominator *increases* the size of the estimated standard deviation (s), thus compensating for the underestimation of the population standard deviation (σ) cited in Number 3 on the opposite page.

Table 7-2 lists the key concepts of this chapter and organizes them for easy access, along with their symbols and defining formulas. The distinctions implicit in that table are essential to a thorough understanding of the transition from descriptive to inferential statistics. No calculation boxes are included in this chapter because you have already calculated all of those measures, with the exception of the population standard deviation as estimated from sample data (s), and that one is exactly the same as the S that you have calculated (see Box 4-1, page 41), except for the substitution of $n-1$ for n.

8

Precision of Inference

When Joe Cawledge obtains a verbal test score of 30, we report his raw score, his standard score, or one or more percentiles. If someone wanted to know the mean score of a group of sophomores in a psychology experiment, we would have the same options there. Of course, neither raw score by itself would have any meaning, but what about the other two (standard score and percentile)? You learned in Chapter 5 to interpret them, didn't you?

Well, yes and no. We have briefly discussed the problem of generalizing from a sample: the question of whether it is appropriate, once we have measured a sample, to make statements about some population that we say the sample represents. Really, though, that is not an either–or question. A proper answer would have to indicate the level of confidence that we have in that answer, and the probable *error* of such an inference is often as important to know as the inference itself.

This chapter, then, is all about errors.[1] Its title is not misleading, however, for it is by discovering the probable limits of error that we define the precision of our inferences. In many applications, that definition is actually more important than the score. There is no point in estimating parameters unless our estimates are *reliable*.

Standard Errors

The concepts "sampling distribution," "sampling error," and "standard error" are closely related. A *sampling distribution* is a distribution comprising estimates (from measurements of samples) of the magnitude of some property of a population; *sampling error* refers to the dispersion of those estimates; and the *standard error* is a measure of that dispersion—namely, the standard deviation of the sampling distribution.

An example will make this clear.

The Standard Error of the Mean ($s_{\overline{X}}$)

One of the most important properties of a sample mean is that in a normally distributed population it is the most stable measure of central tendency. By "most stable" I mean that it varies least among repeated random samples taken from the same population. Let us say we are interested in finding the mean Wechsler Adult Intelligence Scale IQ of American college students. For simplicity, let us assume that the population itself remains stable during the time that it takes us to sample it. Such a population would probably be distributed normally with respect to IQ scores; let's assume that it is. It is too large to be described, so we must draw a sample and infer from it the mean IQ of the population.

There are more ways than one of selecting a sample. Conceptually, the simplest method is *random* selection, in which the composition of the resulting sample is entirely a matter of chance. The ideal case would be the equivalent of picking identification numbers out of a lottery bowl: Each number would represent one student; the entire bowlful, all students. Then we would test all the persons whose numbers had been drawn (our sample of the population with respect to IQ).

Now back to the stability of the mean. Again I ask that you imagine something that would never happen: If we were to select, say, 10,000 individuals in the manner described above, we could compute a mean of their IQs. If we were to return that 10,000 to the population and then select another sample of the same number in the same manner, we could compute a second mean. Returning those subjects to the population, we could select a third sample, and a fourth, and so on ad infinitum. Each sample would have a mean, and not all of those means would be the same. In fact, they would form a distribution—a distribution having the same bell shape as that of the population and of each of the samples. It would, however, be a much more compact distribution than any distribution of individuals. Figure 8-1 depicts distributions of the population, of a single sample, and of the means of many samples on a common scale of IQs.[2]

In Figure 8-1, a common symbol for the mean of the means ($\mu_{\overline{X}}$, pronounced "mew sub ecks bar") has been placed at the mean of the population (μ_X, or "mew sub ecks"), and indeed if an infinite number of means were taken, their mean would be the same as the population mean. But, of course, since that is never done, we can never be sure precisely where the population mean really is. (Remember, we know the population only through the samples that we draw from it.)

What we *can* do is discover the *limits* within which the population mean probably lies. The horizontally compact configuration in Figure 8-1 is the distribution of sample means; *when it is narrow,* we know that *our sample mean is near the population mean,* because when that distribution is narrow, *every* sample mean is near the population mean.

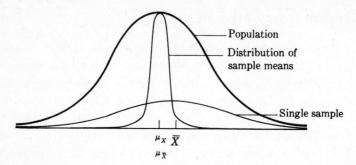

Figure 8-1 Superimposed distributions of the population, a sample, and an infinite number of sample means. μ_X = mean of the population (hypothetical); $\overline{X}$ = mean of one sample (obtained); and $\mu_{\overline{X}}$ = mean of an infinite number of sample means (hypothetical). The mean of the population is an example of a parameter; the mean of a sample is an example of a statistic. Note that the mean of the infinite number of sample means is the same as the mean of the population.[3]

So we have found that which we sought—pinned it down precisely! Unfortunately, however, it is only in fancy that we can take a very large number of samples. In real life, we get only one; so how does all this really help? It helps because although a knowledge of the sample mean will never tell us what the population mean is, we can *estimate* from the variability and the size of the sample what the variability of a distribution of sample means would be. Our estimate of the standard deviation of a hypothetical distribution of sample means is called a *standard error of the mean,* and its defining formula does include both the variability and the size of the sample:

$$s_{\overline{X}} = \frac{s}{\sqrt{n}} \tag{8-1}$$

where $s_{\overline{X}}$ is the estimated standard error of the mean, s is the estimated standard deviation of the population, and n is the size of the one sample that you have observed. From here on, however, I'll refer simply to "the standard error of the mean" without saying "estimated," because you will use it only when you lack access to the entire population of interest, and you always lack that access when you are using a standard error. If you *had* that access, you wouldn't *need* a standard error, because there would *be* no error.

Formula (8-1) implies that the standard error is *small* when the sample is characterized by *small* variability in a *large* number of scores. A small standard error implies that samples of this size and variability have means that tend to be very close to each other. That in turn implies that the mean of our sample probably does not differ very much from the means of other samples that we might

---

Box 8-1   Calculation of Standard Error of the Mean for the
Collegiate IQs Example

$$s_{\overline{X}} = \frac{s}{\sqrt{n}}$$

$$s_{\overline{X}} = \frac{10.2}{\sqrt{10,000}} = \frac{10.2}{100} = 0.102$$

To calculate the *standard error of the mean*, all you need is the *size of the sample* and the *estimated standard deviation of the population*. In this case $n = 10,000$ and $s = 10.2$.

$$n = 10,000$$

$$s = 10.2$$

And of course you can't calculate that standard deviation unless you know the *mean of the sample*:

$$\overline{X} = 115.5$$

---

have taken from the population "collegiate IQs." It means that our obtained statistic is *reliable*.

Since you already know how to compute the estimated standard deviation of the population (s), the calculation of the standard error of a mean is straightforward, as you can see by referring to Box 8-1.

Other Types of Standard Error

Throughout the foregoing discussion of standard errors, actually only one kind—the standard error of the mean—has been presented. That limitation was imposed because, although standard error may be a difficult concept to grasp initially, once it is understood in one situation it can be readily applied to others. The same logic applies to the standard error of a standard deviation, of an obtained score, of a difference between means (Chapter 9), of a correlation coefficient, and many others. In every case, a *small* standard error tells us that the sample statistic is a reliable estimate of the corresponding population parameter.

Confidence Intervals and Levels of Confidence

We now have a powerful tool to use in our attempt to locate the population mean, because now we can try some hypotheses and discover the probabilities of their being valid. Figure 8-2 indicates graphically how that might be done in the case of the mean IQ of college students if the standard error of the mean were $s_{\bar{x}} = 0.102$ (as it would be if calculated from the sample statistics listed in Box 8-1).

Essence of the Concept

The obtained mean IQ is 115.5. Figure 8-2 shows how that statistic plus the standard error of the mean can be used to answer the question, "Within what

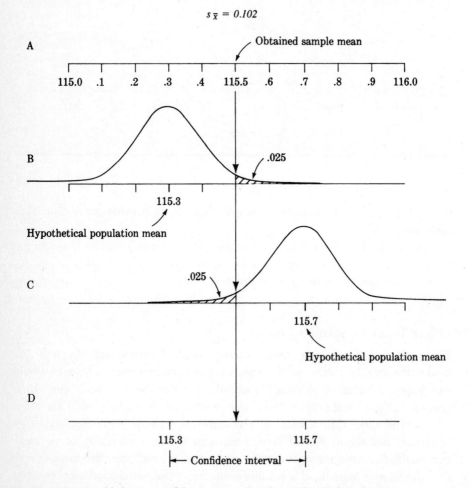

Figure 8-2 Establishing a confidence interval at the .95 level of confidence.

limits may we be reasonably confident that the population mean resides?" The interval enclosed by those limits is called a *confidence interval,* and the "reasonably confident" in the above question is expressed quantitatively as a *level of confidence.* Each of these important concepts depends on the other for its meaning. If I were to tell you that I am 100 percent sure that the true mean is merely *somewhere,* what would you learn about its location? And what information would you acquire if I were to say that the true mean just *may* be between 115.3 and 115.7? You have to know the interval if the level is to mean anything, and vice versa.

Actually, the confidence interval and the level of confidence are directly proportional to each other. Given a small standard error of the mean, we can say with *high* confidence that the true mean is within a *large* interval; but reducing the size of the interval reduces our confidence that the true mean is within that reduced interval. A moment's thought will convince you that this must be so in all cases.

You will recall that there is a variance in the means of random samples from the same population. That being the case, if the population mean were, say, 0.2 of a point *below* the obtained sample mean (Figure 8-2B), how many of a thousand sample means would be as *high* as the one we got? In the drawing, you can see that .025, or 25 out of every 1000, sample means would be that high. We can now state that the population mean is not lower than 115.3, and we can state it with a level of confidence of $1.000 - .025 = .975$.

The lower limit is only half of what we need to define a confidence interval, but with that behind us, the rest is easy. All we do is to repeat the preceding process, except this time we test the hypothesis that the population mean is *above* the obtained mean. Now we ask, "If the population mean were 0.2 of a point *above* the obtained sample mean (Figure 8-2C), what is the probability that a sample mean as *small* as ours would occur?" Again the probability is .025.

Testing our first hypothesis informed us that there is a probability of only .025 that the obtained mean is lower than 115.3; the second test yielded a probability of .025 that it is higher than 115.7. Now let us put the two hypotheses together: "What is the probability that the population mean lies outside the interval 115.3 to 115.7?" The answer is, of course, $.025 + .025$, or .05. So if we say that the population mean *is* somewhere within that interval, we can do so at a *confidence level* of $1.0 - .05 = .95$, and we can speak of it as the 95 *percent confidence interval.*

Some Exercises to Consolidate the Concept

Estimating the amount of deviation of a particular sample mean from a hypothetical population mean is somewhat arbitrary; it depends on where you imagine the true mean to be. But the *shape* of the distribution of sample means in

Figure 8-2 is *not* arbitrary; it is estimated from sample size and variability by computing the standard error of the mean.

The deviation of an obtained mean from a population mean is a deviation score; the standard error of the mean $(s_{\bar{x}})$ is essentially a standard deviation. A deviation score divided by a standard deviation is a standard score. So when we divide our imagined deviation by our hypothetical standard deviation, we obtain a standard score. In Figure 8-2B, the deviation score is 0.2 and the standard deviation is 0.102; the resulting standard score is 0.2/0.102 = 1.96. If we were to consult a table for a normal curve with a standard score of 1.96, we would find that the area in the smaller portion of the curve is indeed .025. If you were learning to *do* inferential statistics, there would be such a table at the back of this book, and you would get much practice in using it. However, since your objective is to *understand* inferential statistics, it is better that you deal with such problems graphically. The following exercises will give you several opportunities to do that. Don't skip over them. On the other hand, don't be too concerned about precision; just do what *looks* right. That will be close enough for the purpose at hand. However, if you think you need help in estimating areas under a normal curve, turn back to Figure 5-4, page 56.

Trace the curve in Figure 8-2 on a piece of paper, take a pair of scissors, and cut around it so that you have a distribution that can be moved about in order to test various hypotheses. Try the limits of 115.4 to 115.6, 115.2 to 115.8, 115.1 to 115.9, and 115.0 to 116.0; estimate in each case the probability that the obtained mean is *outside* the interval and then calculate the probability that it is inside. You can do that for any interval you wish, but in a real-life situation, you would decide in advance how much risk of error you were willing to accept and conversely what confidence level you would require. Then you would find the score limits that corresponded to that level of confidence.

In other words, you would do just what I did in Figure 8-2. I initially decided that I wanted to illustrate a confidence level of .95, which leaves a probability of .05 that the obtained mean lies *outside* the confidence interval. Half of that .05 (i.e., .025) is found above and half below that interval, so I proceeded as follows:

1. First, I slid the hypothetical distribution of means downward from 115.5 until the vertical line from the obtained mean cut off what appeared to me to be about .025 of the distribution, thus designating the mean of that distribution as the lower limit of my confidence interval. If the population mean were any lower than that, fewer than 0.25 of repeated samples would be expected to have means as high as the one in my sample.

2. Having established the lower limit, I repeated the procedure in order to find the upper. I asked how high the distribution of sample means had to be to put only .025 of them below the obtained mean of my sample.

3. Having found that, I had both limits of the confidence interval that corresponded to a confidence level of .95: that is, $1.00 - .025 - .025$.

Effect of *n* on Standard Error

We have already noted that the variability of a hypothetical distribution of means $(s_{\bar{x}})$ is estimated from the variability of an actual distribution of individual cases (S). But another factor is extremely important in making that estimate: the *number* of individual scores (n) in the sample.

The relation of n to $s_{\bar{x}}$ may be less comprehensible initially than that of s to $s_{\bar{x}}$, since the latter relation is between two forms of the same concept; however, the relation to n can be quickly illuminated by means of a few simple diagrams. Figure 8-3A represents the actual distribution of an entire population of scores;

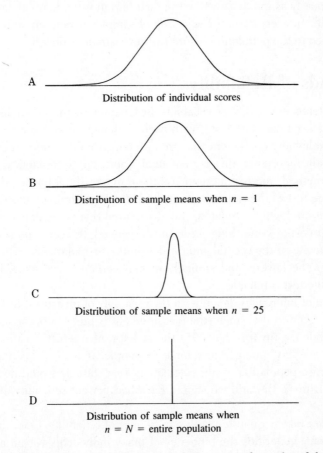

A — Distribution of individual scores

B — Distribution of sample means when $n = 1$

C — Distribution of sample means when $n = 25$

D — Distribution of sample means when $n = N$ = entire population

Figure 8-3 Distribution of individuals and of the means of samples of three different sizes.

each unit in the distribution is an individual member of that population. The three drawings below it are hypothetical distributions; each unit within each of them is the mean of a sample from the above population of individuals. The differences among the three distributions are striking—and those differences are a function of differences in n.

You may have been surprised to see that the first of the hypothetical distributions of means (Figure 8-3B) was exactly the same as the original distribution of individual scores. But a moment's reflection will convince you that it could be no other way, because when $n = 1$, every sample *is* an individual score. Similarly, you can see that if every sample included the entire population (Figure 8-3D), the variability among sample means would have to be precisely zero (every mean would be exactly the same as every other), and the statistic $\overline{X}$ (which is here equal to μ) would be *completely reliable*. Figure 8-3C represents an intermediate situation in which every sample has an n of 25; there you can see some variability, but not nearly as much as where the distribution is made up of samples of 1 (Figure 8-3B). In every case, the selection of samples is random, and each sample is returned to the population before the next sample is drawn.

Two Kinds of Reliability

You encountered one index of reliability in Chapter 6: the coefficient r_{xx} that emerges when you correlate test X with itself. Although there are various ways of obtaining a reliability coefficient that are different from the method cited there, that one well represents the correlational approach to reliability. It is the test–retest method: You test a group and then, maybe the following day, administer the same test to the same group again. If the two administrations yield sets of scores that are highly correlated, you are assured that you will get very similar results if you do the same thing yet again: In general, the same persons will get the highest scores in the second and third as in the first administration, the same ones will get the lowest, and similarly in between. You are assured, in other words, that the test is reliable.

Precision of inferences depends on the reliability of measures. In Figure 8-2, the confidence interval is small (the measurement is precise) because at a given level of confidence (in this case .95) the statistic of interest (in this case, the mean) does not vary much from sample to sample. If we had been able to include the entire population in our calculation, that statistic would not have varied at all (Figure 8-3D) and we should have had perfect reliability and perfect precision.

So you are now acquainted with two indices of reliability. One, the correlation coefficient, quantifies the tendency of many individuals within a specified group to be related to each other in the same way across successive measurements; the other, the standard error, quantifies the sampling error of a single

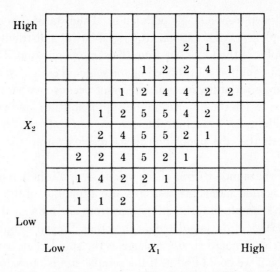

Figure 8-4 Scatterplot of two successive administrations of a test, using digits to represent scores that may be stacked atop one another so that only the top one can be seen. The digit in each cell represents the number of persons whose scores are stacked in that cell. The correlation is $r_{xx} = .75$, which is rather low for a reliability coefficient.

statistic. The two indices are therefore different. On the other hand, they share the essence of reliability: a similarity of results over successive trials. That similarity can be seen graphically as (1) the compactness of a scatterplot of individual scores on two successive administrations of the same test (see Figure 8-4) and (2) the compactness of a distribution of sample statistics (see Figures 8-2 and 8-3).

SUMMARY

Every measure is derived from a sample, and although samples are by definition representative of the population from which they are drawn, they differ from each other even when drawn from the same population. If there are no *systematic* (biased) deviations of the characteristics of the sample from those of the parent population, we say that the deviations are *random,* and we refer to them collectively as *error variance.*

Now, if there is going to be error in our measurement, it behooves us to know its magnitude—or rather, its *probable* magnitude—for we might otherwise be overconfident in the statements that we make about the population from which our sample has been drawn. The most common way of quantifying error is to calculate a *standard error* for whatever statistic we wish to cite.

A standard error is an estimate of the standard deviation of a hypothetical distribution of values that would be obtained for a given statistic if repeated samples were drawn from a single population. If the statistic in question were a mean, for example, we could estimate the variability of a distribution of means of successive samples of the same size from the same population. The standard deviation of that distribution is estimated by the *standard error of the mean*.

The standard error of the mean is important because with it we can mark off the limits within which the population mean probably lies, and we can ascertain what that probability is. In technical terms, we can state the *level of confidence* of our hypothesis that the population mean resides within the specified *confidence interval*.

The size of the standard error of the mean ($s_{\bar{x}}$) is a function of the variability of the sample, as indicated above. But it is also a function of the *number of individuals (n)* in the sample; n varies all the way from *one* individual at one extreme to *all the individuals in the population* at the other. If each sample includes only one individual, the standard error is as large as the standard deviation of the population would be if we could find it. If the sample size is the same as that of the population, the standard error is zero, because repeated samples would all have the same mean. Other sample sizes have intermediate effects on the standard error.

The standard error of the mean has been used here as an illustration of the concept of *sampling error*. Everything that has been said about that here can also be said of the standard errors of other statistics.

The standard error is one kind of *reliability*. Another is the correlation of a test with itself that you first encountered in Chapter 6.

Sample Applications

EDUCATION

You are superintendent of a large urban school district. You want to know the average achievement levels of elementary school children in each grade. Funds are limited, so you test a random sample of 10 percent of the students at each grade level, using a battery of nationally standardized achievement tests. From that information you calculate the mean achievement level of the sample at each grade level. What might you want to know besides the mean?

POLITICAL SCIENCE

You have developed an Index of Political Participation and, as a part of the standardization procedure, have applied it to each of several middle-class neighborhoods. You report the mean of the scores (50 points) to potential users of the index, but since no measure is perfectly reliable, you also want to report the limits within which the *true* mean probably lies. How do you proceed?

PSYCHOLOGY

You have administered an inkblot test to a 10-year-old boy. The results indicate that the boy is mildly disturbed and in need of psychotherapy. Since you know, however, that the inkblot measure is not perfectly reliable (i.e., the results may vary from one administration to the next), you wonder whether psychotherapy should be recommended. (The next administration of the inkblot test might indicate that the child is well within the normal limits.) The test manual might contain something that would help you to decide how much confidence to place in the test score. What would you look for in the manual?

SOCIAL WORK

A family service agency utilizes a Family Life Involvement Profile (FLIP) to measure a family's psychological functioning. The scale yields a score that indicates whether a family's functioning is inadequate, marginal, or adequate. In your assessment of a family, you obtain a FLIP score that indicates marginal functioning. But you wonder how accurate the score is—how much confidence you can have in the rating. How can you quantify that uncertainty?

SOCIOLOGY

A member of your research class notes that you have only a sample of professors in the study of authoritarianism cited in the Chapter 5 sociology application (page 60). He asserts that since you have only a sample, you do not really know where the true population mean is and consequently your conclusions are meaningless. How do you respond?

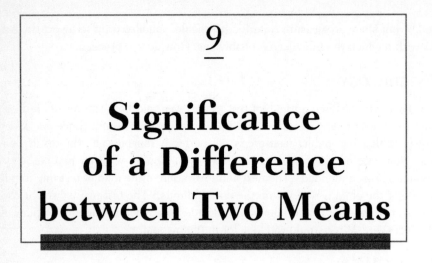

9

Significance of a Difference between Two Means

Often, in both basic and applied research, it is important to know whether two populations are different from each other. From the point of view of the researcher, the question is better stated in the negative: *Are the two samples that I have measured merely two random samples from the same population?* The true answer to that question is either yes or no, but one can never know with absolute certainty which it is. One must therefore state one's answer in probabilistic terms. The probability that the true answer is yes (that the two samples have been drawn from a single population) depends on two factors: (1) the size and direction of the obtained difference between the two means and (2) the variability of a hypothetical distribution of differences between means when pairs of samples are taken at random from a single population. Note the effects of these two factors as you read through the chapter.

One way to illustrate the effect of variability is to analyze carefully a single example. In the following section, we shall examine an experimental study in the field of education, from the design of the experiment to the announcement of the results, concentrating throughout on basic ideas rather than computations.

An Example

Imagine that we have invented a new—and we hope better—method of teaching French grammar to American high school students who have no knowledge of that language. Is it really better than the prevailing method? To find out, we take two samples from the population "naive American high school students," make sure that the two are initially random samples from a single population, and teach one group (hereafter to be known as the *control group*) by the traditional method and the other (the *experimental group*) by the new method. Then, after 150 hours of teaching, we test both groups and compare their mean scores. The objective of our study is to ascertain whether the two are *still* random samples of the same population insofar as their knowledge of French grammar is concerned.

Say the difference between the means of the two groups is 10 points. Is the obtained difference significant? Is it large enough that we may reject the hypothesis that the control and experimental groups remain, after the teaching as they had been before, random samples from a single population with respect to knowledge of French grammar? That single population might now be labeled "American high school students who have had 150 hours of instruction in French grammar." Our hope is that, instead, there are *two* populations—one superior to the other with respect to knowledge of French grammar—and that the experimental group represents the superior population.

This example is admittedly very abstract, because you have to imagine a population that does not exist—that is, a population made up of an infinite number of American high school students who have been taught French grammar by the new method. We suspect that such a hypothetical population would be superior to the one with which we are more familiar, but before we can be sure of that, we must disprove the hypothesis that the two groups are merely two samples of a single population. That is the hypothesis of no difference, or the *null hypothesis* (abbreviated H_0), and our test of significance is really a test of the null hypothesis: We attempt to *dis*prove the hypothesis that there is no real difference between the groups—that the observed difference is merely a chance difference resulting from ordinary sampling error. (See Figure 9-1.)

That may seem easy, but there is a complication. Like virtually every conclusion that emanates from statistical reasoning, this one must be stated not in absolute terms, but in terms of probability. We will not get a yes or a no out of our statistical test. Rather, it will give us *the probability that we are rejecting a true null hypothesis,* and although that probability could conceivably approach zero, we'll never get a flat-out no.[*]

We therefore adopt arbitrarily some level of probability that is an *acceptable* approximation of zero, and when a test reveals a probability below that level, we reject the null hypothesis. (After that, we may entertain other hypotheses.) Whether a particular difference—like the 10 points we obtained between our control and experimental groups after the language instruction—meets that

---

[*]If we do not reject the null hypothesis, we must regard it as tenable (our data offer insufficient evidence against it) but not necessarily true.

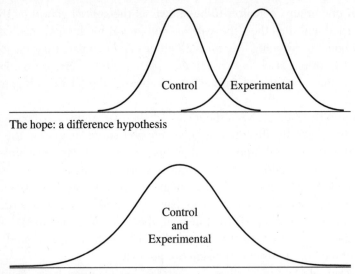

The hope: a difference hypothesis

The test: a no-difference hypothesis

Figure 9-1   Since we have conceived the new method as an improvement on the old, we hope that its superiority will be manifest when the final tests are scored. But to demonstrate that an obtained difference between the two groups was not obtained by chance, we must test the hypothesis that students who constitute *two* groups with respect to teaching methods are nonetheless *one* with respect to performance. That is a hypothesis of no difference—the null hypothesis (H_0).

preset criterion depends upon the *variability of a hypothetical distribution* in much the same way that the size of the confidence interval did when we were discussing the stability of the mean. It might be worth your while to review that section before proceeding with this one; it begins on page 97.

Before attempting to quantify the probability that a given difference has occurred strictly by chance, it might be helpful to examine Figure 9-2 and react to it intuitively. You can see that in each pair of distributions—pair A and pair B—sample means are distributed symmetrically about two points and that the distance (difference) between these points is the same in pair B as it is in pair A. But are the two differences equally impressive? Which is the more reliable? . . . Clearly the one in pair B, isn't it? The two distributions with the smaller standard errors (pair B) are the more reliable because there the means vary only a little from sample to sample of their target populations. That makes the *difference* between B1 and B2 more reliable, too, than that between A1 and A2.

Now to approach the same idea more analytically, look at Figure 9-2 again and imagine taking one mean from distribution A1 and one from A2, calculating the difference between them, repeating that procedure until every mean has a counterpart in the other distribution, and then doing the same with distributions

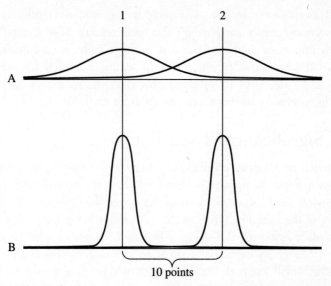

Figure 9-2 When individual scores vary only a little around their respective means, any difference between these means is reliable. That condition obtains in diagram B much more than it does in diagram A.

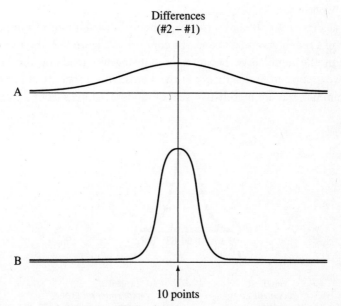

Figure 9-3 The *reliability* of a difference between means viewed as the *compactness of a distribution* of differences between means. Distribution B is more compact than distribution A. See text for procedure.

B1 and B2. Because the location of sample means is more reliable in B than in A, the *differences between* samples are also more reliable. (See Figure 9-3.) That is extremely important right now, because in this chapter we are dealing primarily with a hypothetical distribution not of means but of *differences between* means. In this case, each difference is between the means of two groups of American high school students taught by different methods.

Test of Significance: The z Ratio

Is a difference of 10 points sufficiently large to be significant, or is it small enough that it might easily have occurred by chance—by ordinary sampling error? The answer is a function not only of the size of the difference itself but, very importantly, of the sampling errors of the two groups being tested—as indicated by the standard errors of the means. The next six paragraphs, together with Figures 9-4 through 9-7, concern a situation in which the standard error of the mean is rather small and is the same for both groups. (Just accept my figures for the standard errors throughout this chapter. Concentrate now on what you can *do* with a standard error once you have it.) You may find this discussion difficult to follow the first time through; it would be easier just to learn the formula and its practical implications and be done with it. But the diagrams hold your best hope of really understanding how the standard error of the mean relates to the standard error of a difference between means and how the latter functions in the testing of hypotheses.

Look at Figure 9-4. It shows two hypothetical distributions of sample means on a scale of French grammar scores. In contrast to Figure 9-2, the two distributions occupy the same space because the drawing was made on the hypothesis that the two groups are in reality *not* of two kinds with respect to achievement in French grammar, but are merely two sets of random samples from a single popu-

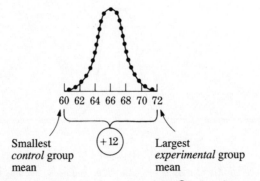

Figure 9-4   Sample means, $s_{\bar{x}} = 2$, when control ($\frown$) and experimental ($\bullet\ \bullet\ \bullet\ \bullet$) populations are identical.

lation. That is, of course, the null hypothesis (H_0), and we shall try our best to disprove it (or rather, to make it untenable).

In order to test the null hypothesis with respect to differences, it is necessary to think in terms of a scale not of scores but of differences between scores—specifically, of differences between control group means and experimental group means. Look again at Figure 9-4 and imagine taking pairs of means (one control and one experimental group in each pair) at random from the two distributions. (Remember that according to the null hypothesis, the two are actually one.) Most of the intrapair differences would be close to zero, but a few— just by chance, remember—would be substantially larger. In fact, if you were to compare the largest experimental group mean with the smallest mean of the controls in this example, the difference would be 12 raw-score points, or 6 standard errors in the distribution of means (Figure 9-4).

That difference, $\overline{X}_e - \overline{X}_c$, is in the direction we have predicted, so we'll call it positive. But if all the samples are drawn from the same population, as the null hypothesis would have it, there should be as many negative differences as there are positive, and the largest of them should be just as impressive as the positive difference we just found. Figure 9-5 represents the same two distributions as Figure 9-4; if you will compare the *lowest* experimental with the *highest* control group mean, you will indeed discover that the difference is again 12, but this time in the negative direction.

Figures 9-4 and 9-5, then, represent the same pair of distributions analyzed in two different ways: one as positive, the other as negative differences. Figure 9-6 combines positive and negative differences into a distribution of differences. Positives are on the right, negatives are on the left, and the mean is zero—the null hypothesis again. If we were to compute an infinite number of differences between pairs of means, selecting each pair at random from a single population, that is about what the distribution would look like.

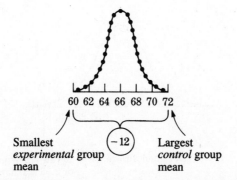

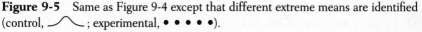

Figure 9-5 Same as Figure 9-4 except that different extreme means are identified (control, ⌢‿⌢ ; experimental, • • • • •).

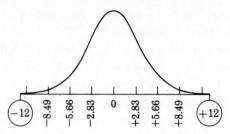

Figure 9-6   Differences between means, $s_{\bar{X}_e - \bar{X}_c} = 2.83$, on a scale of raw scores.

With that much behind us, the testing of the null hypothesis should be relatively easy. All we have to do is place our obtained difference within the hypothetical distribution of differences, as in Figure 9-7, and we can see immediately that only a very few positive differences of that magnitude (10 or larger) occur simply by chance among samples taken from a single population. (Asking only about *positive* differences constitutes a *one-tail test*. You will find a discussion of one- versus two-tail tests beginning on page 120.)

By inspection of Figure 9-7 it appears that very few such differences would be as large as the one we obtained in our experiment if the samples were from the same population. We are justified, therefore, in rejecting the null hypothesis (H_0) as untenable. Formula (9-1) offers a more precise estimate of the variability of differences between means[1]

$$s_{\bar{X}_e - \bar{X}_c} = \sqrt{s_{\bar{X}_c}^2 + s_{\bar{X}_e}^2} \tag{9-1}$$

where $s_{\bar{X}_e - \bar{X}_c}$ is the standard error of the difference between means,* $s_{\bar{X}_c}$ is the standard error of the mean for the control group, and $s_{\bar{X}}$ is the standard error of

*Again, as with the standard error of the mean, this standard error is necessarily "estimated."

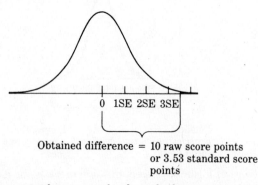

Obtained difference = 10 raw score points
or 3.53 standard score
points

Figure 9-7   Same as Figure 9-6 but on a scale of standard error units. ($s_{\bar{X}_e - \bar{X}_c}$). The shaded area is so small that it is hard to see.

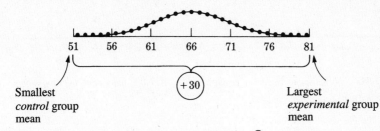

Figure 9-8   Sample means, $s_{\bar{X}} = 5$, when control (‿⌒‿) and experimental
(• • • •) populations are identical. Compare with Figure 9-4.

the mean for the experimental group. Since our interest is in the underlying logic of the significance test rather than in the precision afforded by the formula, we need only note that, like Figures 9-4 through 9-7, the formula shows that the variability of random differences between means taken in pairs ($s_{\bar{X}_e} - s_{\bar{X}_c}$) is proportional to variability among means taken singly ($s_{\bar{X}_c}$ and $s_{\bar{X}_e}$).

To illustrate that point further and to emphasize the importance of $s_{\bar{X}}$ in determining the fate of the null hypothesis, look at Figures 9-8 through 9-11 and see how our 10-point obtained difference would have fared had the sample means been less stable. In Figures 9-8 and 9-9, the standard error of the mean is 5 (instead of the $s_{\bar{X}} = 2$ of Figures 9-4 and 9-5); and in Figures 9-10 and 9-11, the standard error of the difference is 7.07 (instead of the 2.83 of Figures 9-6 and 9-7). Whereas in Figure 9-7 our difference was 3.53 standard errors ($z = 3.53$) above the mean and had a probability of .0002, in Figure 9-11 a difference of 10 is only 1.41 standard errors away from the mean and has a probability of .0793.

So Figure 9-11 represents circumstances in which our obtained 10-point difference is unreliable. (If we were to take 1000 pairs of samples at random from a single population, 79 of the intrapair differences would be at least as large as ours and in the same direction.) Had we obtained our difference in the circumstances represented by Figure 9-11, we would have had to accept the null

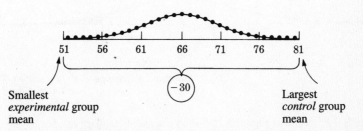

Figure 9-9   Same as Figure 9-8 except that different extreme means are identified
(control, ‿⌒‿ ; experimental, • • • •). Compare with Figure 9-5.

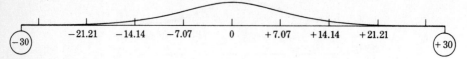

Figure 9-10   Differences between means, $s_{\bar{X}_e - \bar{X}_c} = 7.07$, on a scale of raw scores. Compare with Figure 9-6.

hypothesis as tenable. Incidentally, do not be disturbed if you cannot estimate the above values with this kind of precision simply by looking at the graphs; I did it by formula.[2]

   In summary, Figures 9-4 through 9-7 illustrate a test of the significance of a difference when $s_{\bar{X}_c}$ and $s_{\bar{X}_e}$ are small, and Figures 9-8 through 9-11 test that same difference when those standard errors are large. Distributions are all hypothetical. All the examples in this section have been graphed in order to expose the structure of their operations. Normally, however, those operations are done algebraically for greater accuracy. In the language-teaching study, the formula would be

$$ z = \frac{(\bar{X}_e - \bar{X}_c) - 0}{s_{\bar{X}_e - \bar{X}_c}} \tag{9-2} $$

where z is the ratio of an *obtained difference* $(\bar{X}_e - \bar{X}_c)$ to the standard error of the difference between means $(s_{\bar{X}_e - \bar{X}_c})$ in a distribution of such differences when the mean of the distribution is zero. In Figure 9-7, that ratio is 10/2.83, which makes it a much higher ratio than the 10/7.07 of Figure 9-11. The zero in the formula has no effect on the outcome; it is there only to remind you of precisely what the formula is supposed to do: namely, to ascertain how far our obtained difference is from the mean of a distribution of differences, all of which are the results of sampling errors. The mean of such a distribution would indeed be zero.[3]

   Box 9-1 shows how a z could be calculated from the defining formula (9-2).

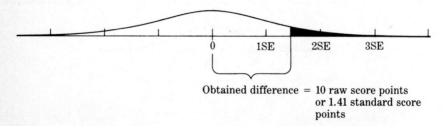

Obtained difference = 10 raw score points
or 1.41 standard score
points

Figure 9-11   Same as Figure 9-10 but on a scale of standard error units $(s_{\bar{X}_e - \bar{X}_c})$. Compare with Figure 9-7; here the shaded area is much larger.

BOX 9-1   Calculation of a Test of Significance of a Difference between Two Means [See Equation (9-2) and Figures 9-4, 9-5, 9-6, and 9-7]

(1) $\overline{X}_e - \overline{X}_c = 10$

(2) $s_{\overline{X}_e} = 2 \qquad s_{\overline{X}_c} = 2$

(3) $s_{\overline{X}_e - \overline{X}_c} = \sqrt{s_{\overline{X}_c}^2 + s_{\overline{X}_e}^2} = \sqrt{2^2 + 2^2} = \sqrt{4 + 4}$

$\qquad\qquad = \sqrt{2^2 + 2^2} = \sqrt{4 + 4} = \sqrt{8} = 2.83$

$$z = \frac{(\overline{X}_e - \overline{X}_c) - 0}{s_{\overline{X}_e - \overline{X}_c}} = \frac{10}{2.83} = 3.53$$

If there are 50 students in the control group and 50 in the experimental, $p \leq .01$.

Before you can calculate the ratio z of a difference between means to the standard error of a difference between means, you must:

measure all the individuals in both samples and calculate two means,

calculate the difference between the means (row 1),

calculate the standard error of the mean for each sample (row 2), and

calculate the standard error of the difference between means (row 3).

The ratio of row 1 to row 3 is:

$$\frac{\text{row 1}}{\text{row 3}} = \frac{\text{difference between means}}{\text{standard error of difference between means}} = z$$

Test of Significance: The *t* Ratio

There are two important points to be made about Formula (9-2): First, this ratio is of the same form as the one in Formula (5-1), page 50, so that you can apply here what you learned there. There we transformed a deviation score into a standard score by dividing it by the standard deviation of a sample; here we transform a difference between means into a standard score by dividing it by the standard deviation of a hypothetical distribution of differences between means. The second point to be made about Formula (9-2) is that although the general form of this new ratio is the same as that of Formula (5-1), its content is somewhat different: Specifically, the distribution here is hypothetical, not actual, and the z ratio can be used only when the distribution is actual—that is, when we can measure every member of the population. Because measuring every person in an entire population is not always feasible, we need a ratio that will work when we can measure only a *sample* of the population.

We may regard Equation (9-2) as the basic formula for the significance test because it clearly reveals the relation between the difference and the standard error of the difference. Unless you have measured the entire population, however, it is not the z ratio that we will test for significance, but a ratio called t. The t ratio is essentially the same as z, but it is referred for significance testing to a distribution that changes its shape (because the probability of obtaining the various scores in it changes) as n gets smaller. When n is very large, the difference between z and t distributions is negligible; when n is as large as the population, the two are identical.

Because t was developed specifically for use with small samples, it is important to use degrees of freedom (df) in place of n when discussing it (see pages 93–94). In Figure 9-12, it is clear that when the t distribution is used, the probabilities to be inferred from various placements on the baseline are in many instances quite different if degrees of freedom is small than if it is large. Most notably, when degrees of freedom is small, extremely large t ratios (either positive or negative) make up a larger-than-normal part of the distribution of samples.

Because it is appropriate for use with either large or small samples, the t test is used almost universally in place of z whenever inferences must be made from accessible samples to inaccessible populations. But because z is a construct that you already understand, it is best to think of t as a modified z.

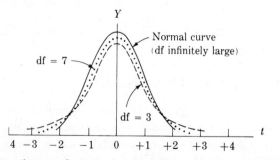

Figure 9-12   Distribution of t ratios at three different sample sizes. [Adapted from Henry L. Alder and Edward B. Roessler, *Introduction to Probability and Statistics*, 6th ed., W. H. Freeman and Company. Copyright © 1977.]

Significance Levels

We have seen in Figures 9-8 through 9-11 that in certain circumstances a difference of 10 points between two sample means can occur frequently by chance.
However, our difference was obtained not in the circumstances depicted in
Figures 9-8 through 9-11, but in those of Figures 9-4 through 9-7; there it is
clear that the probability is extremely small that, with respect to their knowledge
of French grammar, our control and experimental groups are random samples of
a single population. Now we may announce to an eagerly waiting public that our
new method of teaching French grammar is almost certainly better than the traditional one—at least under the circumstances prevailing in our experiment. We
proclaim that superiority at a *significance level* of .0002, because there is only
that much of a probability that our proclamation is wrong—that we are rejecting
a true null hypothesis.

It may seem to you that the level of significance in this instance should be
1.000 − .0002, or .9998. But when we reject the null hypothesis, convention
decrees that the significance be expressed as *the probability that a true null hypothesis is being rejected.* That means that the *lower* the significance index (in
this case .0002), the *higher* is our confidence that the effect we have observed is
real—that it is *reliable.*[4]

Tradition also provides us with two levels recognized as significant and very
significant, respectively. A *significant* difference is one that would occur only 5
times (or fewer) in 100 comparisons if every sample were taken at random from
the same population; a *very significant* difference is one that would occur only
once in 100 comparisons. These are sometimes referred to as the .05 and .01
levels of significance, or as $p \leq .05$ and $p \leq .01$, respectively. (The p stands for
"probability," the $\leq$ "less than or equal to.") In tables and other places where the
briefest possible abbreviations are used, a significant difference sometimes is
designated simply by "S" and a very significant one by "VS." In practice, those
two levels are often used but also often ignored. For example, a researcher who
obtained as impressive a difference as that between our control and experimental groups in the language-teaching study ($p \leq .0002$) would be unlikely to hide
his or her light under the bushel "less than .01"!

A Common Misinterpretation

Given the information that $p \leq .01$, you may be tempted to say that because
there is at most a .01 probability that a difference as large as the obtained one
occurred by *chance*, there is a probability of at least .99 that the obtained difference is *real.* But the significance test is concerned only with the null hypothesis,
and the null hypothesis asserts that the real difference is *zero.* A p of .01 tells us
only that if the two measured groups are random samples from a single
population—that is, if the null hypothesis is true—then either

1. the probability of getting a difference as large as this (the two-tail test) is no greater than .01, or

2. the probability of getting a difference as large as this *in the expected direction* (the one-tail test) is no greater than .01. In both cases, the "real" difference (on the null hypothesis) is zero.

One- versus Two-Tail Test

I have just used the terms *one-tail test* and *two-tail test*. The term "tail" refers to the upper or lower end of a normal frequency distribution where the curve is close to the baseline. Figure 9-7 is a diagram of a one-tail test. The shaded portion *in the right tail* is very small: Specifically, it represents a .0002 probability that a difference as large as 10 points *and in the expected direction* would occur by chance.

The alternative is a two-tail test, which is appropriate whenever *we have not predicted the direction* of the difference. If you will look back at Figure 9-6 for a moment, you will see that the probability of a difference of 10 in *either* direction is double that of a difference in *one* direction. That means that a particular difference can fail the significance test if there has been no directional prediction and pass it if a prediction has been made (provided, of course, that the outcome matches the prediction). We are not free to select either test arbitrarily, however; if we have no reason to expect a difference in one direction rather than another, we are obliged to use the two-tail test. Only when we do have such a reason—*when we can make a prediction*—and when the results confirm the prediction are we justified in using the one-tail test.[5]

Statistical versus Practical Significance

The length of this section barely exceeds that of a long footnote, but it contains an important notion that might easily be overlooked if it were not emphasized. The point is that even though a difference is shown to be statistically significant, it may not have any *practical* significance.

The result of our language-teaching experiment was very convincing; there can be almost no doubt that the difference between the two groups at the end of the teaching period was real. But how does that information affect the decisions of a school administrator who must decide whether to adopt the new method? That depends on many considerations, some of which have nothing to do with statistics. It depends to a great extent on how costly the new teaching method is to implement—whether it requires expensive equipment or specially trained teachers. The administrator's decision may also depend on the absolute size of the difference. That may seem a contradiction of everything we've been saying up until now, and it is—unless we keep in mind the distinction between statistical and practical significance.

The joker is in that standard error term—more specifically, the importance of n in determining the size of that term. If you analyze either Figures 9-4 to 9-11 or Formula (9-1), you will find that the size of the standard error of the difference is proportional to the standard errors of the means of the two samples. The size of a standard error of the mean [Equation (8-1)] is determined partly by the *variability* in the sample and partly by its *size:*

$$s_{\overline{X}} = \frac{s}{\sqrt{n}}$$

Now, the significance of an obtained difference depends upon the ratio between that difference and the standard error of the difference:

$$z_{\overline{X}_e - \overline{X}_c} = \frac{\overline{X}_e - \overline{X}_c}{s_{\overline{X}_e - \overline{X}_c}} \tag{9-3}$$

where $z_{\overline{X}_e - \overline{X}_c}$ is the difference between two means in standard error units, $\overline{X}_e - \overline{X}_c$ is the difference between two means in raw-score units, and $s_{\overline{X}_e - \overline{X}_c}$ is the standard error of the difference. The presence of n in the denominator of the $s_{\overline{X}}$ formula means that a large n yields a small $s_{\overline{X}}$. That, we know, reduces the standard error of the difference, and because z increases as the standard error decreases, the ultimate result of a large n is high significance. That is why if our samples are large enough, *any* difference will be statistically significant, no matter how small it may be. Thus, although a z ratio of 3.53 indicates that the difference, no matter how small, is probably genuine, a school administrator might decide that a very *small* difference would not be worth its additional cost, even if it were *absolutely* reliable—that is, even if $s_{\overline{X}_e - \overline{X}_c}$ were zero.

The point is that some differences are too small to be of practical significance and that the size of a difference is in no way affected by its reliability. The temptation to infer that it is should be firmly suppressed.

SUMMARY

Sometimes it is important to know whether two groups are different from each other—or, more often, whether they represent *populations* that are different from each other. In this chapter, we have compared the performance of two groups of students exposed to two different teaching methods. One group, taught by the traditional method, was called the *control group;* the other, taught by a new method, was the *experimental group.* The experimental group outscored the control.

But two groups given the *same* treatment could have scored differently by chance, and in such a case we would have to admit that the obtained difference was a *sampling error*—that the two groups are really only two samples from the same population with respect to the performance being tested. We must consider the possibility that our control and experimental groups are like that, too—that the difference we obtained was due entirely to sampling error and that the two groups are really only two samples from a single population. That possibility is known as the hypothesis of no difference, or *null hypothesis* (H_0).

The general question, then, is really "How large does an obtained difference have to be before we are justified in rejecting the null hypothesis?" The procedure by which that question is answered is known as a *test of significance.* In the case at hand, we ask a more specific form of the question, namely, "Is our obtained difference large enough to justify a rejection of the null hypothesis?"

The answer will not be a simple yes or no; it must be given in terms of *probability.* What is the probability that a difference as large as ours would occur if there were no real difference in the effectiveness of the two teaching methods? Acceptable levels of probability are somewhat arbitrary, but two such *levels of confidence* have traditionally been set at .05 and .01. However, a researcher who obtains a null hypothesis probability much lower than .01 may report it exactly, because the lower it is, the more impressive is the outcome of the study.

Is it possible to ask all the above questions about any measurement of any two groups of subjects? We could use the test of significance simply to explore—to search for differences worthy of further investigation. In the case of the two teaching methods described above, however, we were not exploring; we *expected* the experimental group to be better than the control. That is an important distinction, because if we can state our expectancy in advance, we are privileged to use a *one-tail test* instead of a *two-tail test* of significance. If we are exploring, we must ask, "What is the probability of a sampling error this large *in either direction?*", whereas if we have stated our expectancy, we may ask instead, "What is the probability of obtaining this large an error *in favor of the experimental group?*" Since the probability of a difference in *one specified direction* is just half that of the same amount of difference in *either direction,* the one-tail test is twice as sensitive as the one we would be obliged to use if we were merely exploring; a difference half as large will qualify at whatever level of significance we have prescribed. Finally, a test of statistical significance tells us how *reliable* our difference is, but that is only one factor in making practical decisions.

Sample Applications

EDUCATION

You are principal of a high school. The teachers, counselors, and administrators of the school have developed a one-semester group-counseling program for students who are disrupting class to help those students learn more appropriate ways of resolving conflicts and participating in classroom activities. To find out whether the program helps reduce student disruptiveness, you randomly assign half of the 50 most disruptive students to the group-counseling program. At the end of the semester, the teachers rate the disruptiveness of all 50 students. When the scores have been collected, what do you do with them?

POLITICAL SCIENCE

Increasingly in recent years, political scientists have used statistical methods to evaluate the effectiveness of public programs. An example of this kind of research can be seen in the following case involving a crime control program: The citizens of Gritty City have demanded that local officials do something to resolve the problem of burglaries. You are chief of police. As a first step, you propose that the city council establish a special task force that will give presentations in each neighborhood on how to prevent burglary. Since the council is hesitant to fund the task force on a permanent basis without any evidence of its effectiveness in reducing burglaries, its members agree to make a decision after analyzing data from a pilot program. To set up such a program, you draw two random samples from the population comprised of all of the city's precincts. The residents in one group then receive the task force presentations; residents in the other group do not. After three months, you compare the mean number of burglaries in the precincts receiving the presentations with that in the cops-only precincts. What is an appropriate statistic for making such a comparison?

PSYCHOLOGY

As a clinical researcher, you are interested in ascertaining how a period of training in muscle relaxation will compare with the use of stimulant drugs in reducing hyperactivity in young children. You assign half of the children medically classified as hyperactive to the relaxation program and half to the drug program. Following a 30-day period of intervention, the activity level of all children in the study will be assessed by means of a rating scale. There will almost certainly be some difference between the two groups. How can you tell whether the difference is significant?

SOCIAL WORK

As a social worker in a senior citizens' center, you are concerned about the health and vitality of the seniors who frequent the center. It is your assessment that the center's current program of bingo, pool, backgammon, quilting, films, and occasional field trips is not sufficient to keep seniors active and alert, for they experience numerous health problems, including strokes, heart attacks, upper respiratory illnesses, and emotional illnesses involving depression and anxiety. After attending a workshop on services for the elderly, you plan to implement a new program that includes group discussion, meditation, and physical exercises. It is a structured program; seniors meet for two hours twice weekly. To ascertain the effect of this new program, you randomly select half of the seniors and engage them in the program for a year. You then plan to compare this group to the half of the membership that has continued in the regular center activities. After collecting health data on all your subjects, you find a difference in favor of the experimental group. How can you estimate the probability that this difference arose by chance?

SOCIOLOGY

A family planning agency has asked you whether Catholic families are larger than non-Catholic families in your state. You draw a random sample from census data and find that the mean size of Catholic families is indeed larger than that of non-Catholic families. How might you test the significance of this difference?

10

More on the Testing of Hypotheses

A chapter with this title could easily occupy as much space as all the rest of this book put together. It will not, because I have selected just two tests of significance as illustrations. Others will be mentioned only briefly, if at all. For our purposes, it is not necessary to describe those others. In fact, probably the most important contribution that I can make to your understanding of all of them can be stated without *any* illustrations. The fundamental idea is this: Each is conceived as the *testing of a null hypothesis*—a hypothesis of no difference.

The difference that emerged from our experiment with two methods of teaching French grammar was a difference between scores—a difference in *amount*. We asked whether an innovative method resulted in students who were more able than students taught by the more traditional method with which it was being compared. More technically, we asked whether the difference we obtained was attributable to the difference in teaching methods or merely to sampling errors—that is, we tested the null hypothesis.

But we could have asked a different question: Given some clear criterion of passing, does the innovative method produce fewer failures than the traditional? That question is posed in terms of *frequencies*. Indeed, frequencies are often used to indicate magnitudes, as when number of words correct indicates amount of typing skill or number of strokes (or rather its inverse) indicates amount of golfing skill. The significance tests described in Chapter 9 can be used in such cases, and they can be modified to deal with the categorical cases just mentioned (pass–fail, yes–no, innovative–traditional, etc.). More often, however, such cases are analyzed in a different way. The first section of this chapter will examine a test of the null hypothesis in which all the data are in the form of frequencies.

There is a feature of the example used in Chapter 9 that is not characteristic of all experiments: Only two "treatments" were compared. What should we do if we wanted to assess the relative effectiveness of *several* teaching methods? Or what if we suspected that one treatment method might be effective in the hands of one teacher and another might be superior when used by a different teacher? The remaining sections of the chapter will describe a kind of analysis by which both of those questions can be answered.

Notice that every test of significance is a test of a null hypothesis; conversely, any test of the null hypothesis is a test of significance. Since any observed difference can result from sampling errors, any difference can be put to a test of significance.

Comparison of Frequencies: Chi-Square (χ^2)

Measurements that we have discussed so far have been mostly *quantitative:* Test scores have been referred to *scales* that tell us for a given dimension which individuals are higher, which lower, and by how much. There are other, somewhat less refined possibilities, however. One of those is the identification not of an amount of some attribute of an individual but rather a category, or *class,* into which that individual fits. Differences between two individuals can be quantitative or *qualitative.* If one of the individuals being compared is a horse and another an elephant, it is possible to measure each on a dimension of speed, say, or of quickness, or of strength. Those are *quantitative* measurements. To identify one individual as a horse and another as an elephant, on the other hand, is to *classify* both; and that is a *qualitative* measurement.

It may seem to you, as it did to me at first, that the term "measurement" is misapplied to the mere act of identification. (Measurement experts sometimes even refer to the result of such acts as a "scale.") But to classify is to make discriminations, and making discriminations is the basis of *all* measurement. Arranging measurements in order (as we did on pages 62–65) is more refined than mere classification, and discerning the distance from one to another is more refined than either.[1] Nevertheless, qualitative distinctions by themselves can be very useful. You may never have occasion to compare horses with elephants; however, comparisons of individuals in such categories as "man/woman," "Republican/Democrat," or "artist/scientist" might be candidates for your attention.

Given the importance of such distinctions, we need a significance test to deal with categories and the numbers of individual cases that we find in those categories—that is, with *frequencies*—and with *proportions* and *probabilities,* both of which are based on frequencies. We will use such a test in situations that are not amenable to the application of finely divided quantitative scales. In a public opinion poll, for example, respondents are usually asked a question to which they are expected to answer either yes or no, for or against, and so forth. There are ways of obtaining more finely graded responses, but they are more cumbersome to administer than the single question. Furthermore, a relatively crude index is often the most appropriate to the circumstance. For example, in predicting the outcome of an election, every response that a voter makes to his ballot represents what is essentially a yes–no decision.

Let us consider briefly the reasoning behind a significance test of a set of data from an election. Imagine that a particular male candidate for public office has considerably more sex appeal than another but seems to differ very little from his only opponent (another male) on any other dimension. He wins, and we wonder whether his sex appeal played a part in his victory.

If we assume that the voters of both sexes have a basically heterosexual orientation, we may get an informative answer by rephrasing the question: "Did women vote for Mr. Sex in significantly larger numbers than did men?" The data can be arranged in a 2×2 table (Table 10-1). If there are in this election 20,000 voters—10,000 male and 10,000 female—of which we have a 1 percent representative sample, if Mr. Sex has received 60 percent of all the votes cast, and if there is no difference between men and women with respect to voting behavior in this election, we can expect a certain frequency in each cell of the table.

We can calculate cell frequencies once we know (1) the proportion of the 200 persons in our sample who voted for Mr. Sex and the proportion who voted for his opponent and (2) the proportion of participants who are male and the proportion who are female. That information is recorded in the columns and rows totals of Table 10-2. By dividing the 120 "For" votes by the total number in the sample (200), we find that the proportion *for* Mr. Sex is .6; doing the same calculation in the next column (80/200) yields the proportion *against* him (.4). Thus we should expect .6 of both males and females to vote *for* Mr. Sex (and .4 *against*). Similarly, 100 of the 200 participants are *male* and 100 *female,* which means we should expect half of the voters who were *for* Mr. Sex and half of those who voted *against* him to be male (and half female).

By now you may have suspected that we are preparing to state a null hypothesis. Table 10-3 does that in terms of *expected frequencies.* As you move *across*

TABLE 10-1   **Arrangement of data in election problem**

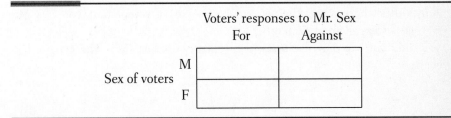

TABLE 10-2   Calculation of expected cell proportions from obtained total frequencies (f) in rows and columns

		Voters' responses to Mr. Sex			
		For	Against	Totals (f)	Proportions of M & F
Sex of voters	M			100	$\frac{100}{200} = .5$ Male
	F			100	$\frac{100}{200} = .5$ Female
Totals (f)		120	80	200	
Proportions of For and Against		$\frac{120}{200}$ $= .6$ For	$\frac{80}{200}$ $= .4$ Against		

the table from left to right, look at the calculation at the *top* of each cell. Moving across, then, you see that .6 of the 100 males in the sample equals 60 male voters; so if the null hypothesis is confirmed, the frequency in the males-voting-for-Mr.-Sex cell will be 60. We say that 60 is the *expected frequency* for that cell. Analysis of the cell to the right of that one proceeds in exactly the same way, but the expected proportion is .4 instead of .6, and the resulting frequency is 40 instead of 60.

Moving *down* the table now, look at the calculation at the *bottom* of each cell. Beginning with males for Mr. Sex, you find that on the null hypothesis, half (.5) of the 120 Mr. Sex voters (60) should be male. And in the cell below that, of course, the other half (60) should be female. Finally, of the 80 persons who voted against Mr. Sex, half should be male and half female.

You will remember that our test of the null hypothesis will consist of a comparison of frequencies *expected*, as calculated in Table 10-3, with frequencies actually *obtained* in the sample as displayed in Table 10-4. Table 10-5 compares the obtained frequencies with the expected frequencies. It is clear that our candidate received a higher proportion of the female vote than of the male. But is the difference *significant*? How likely is it that a difference (actually a set of four differences in Table 10-5) as large as that obtained would occur in a random sample if there were no difference at all within the population of voters? (The null hypothesis here is that the two variables—"sex of voters" and "votes for Mr. Sex"—are *independent* of each other.)

To find out, we may use a statistic called *chi-square* (χ^2). The general formula is

$$\chi^2 = \Sigma \frac{(f_o - f_e)^2}{f_e} \tag{10-1}$$

TABLE 10-3   Expected cell *frequencies* under null hypothesis

| | | Voters' responses to Mr. Sex Expected cell f | | Obtained |
		For	Against	Total f
	M	.6 (100) = 60 \boxed{60} .5 (120) = 60	.4 (100) = 40 \boxed{40} .5 (80) = 40	100
Sex of voters				
	F	.6 (100) = 60 \boxed{60} .5 (120) = 60	.4 (100) = 40 \boxed{40} .5 (80) = 40	100
Obtained Total f		120	80	200

where χ^2 is chi-square, f_o is the obtained frequency, and f_e is the expected frequency. Focus on one part of that formula for a moment

$$f_o - f_e$$

and you will grasp the fundamental nature of chi-square. In each cell of the table, the greater the deviation from the expected frequency, the larger chi-square is likely to be. Notice in Formula (10-1) that squaring each difference allows negative differences to augment rather than reduce the total.

TABLE 10-4   Obtained frequencies from actual sample

| | | Votes for Mr. Sex | | |
		For	Against	Total
	M	50	50	100
Sex of voters	F	70	30	100
	Total	120	80	200

TABLE 10-5   Deviations obtained from expected
frequencies (expected subtracted from
obtained in each cell)

		Votes for Mr. Sex	
		For	Against
Sex of voters	M	50 − 60 − 10	50 − 40 + 10
	F	70 − 60 + 10	30 − 40 − 10

Since the expected frequency is derived from the hypothesis of no difference, chi-square is an index of deviation from that hypothesis. Tables are available in which one can find the significance level of any given chi-square.[2] In the present example, chi-square is 8.34, which is significant at the .01 level; the probability that our obtained difference was due to sampling error is less than 1 percent.

Calculation of chi-square from the defining equation (10-1) is demonstrated for the Mr. Sex election problem in Box 10-1.

Multimean Comparisons: Analysis of Variance

In an experiment, an *independent variable*—for example, a teaching method—is a variable that is manipulated by the experimenter. (Mathematically, it is independent; in an experiment, it is really a *manipulated* variable. But the mathematical term is frequently used whether the context is mathematical or experimental.) For that reason, it is sometimes called a *treatment* variable. The *dependent variable*—for example, student response to teaching—is a variable whose values are determined by those of the independent variable(s) or by others that have been inadequately *controlled*. (A more extended discussion of the variables in an experiment begins on page 151.)

In our evaluation of teaching methods in Chapter 9, we applied a significance test to a difference between two means. That technique is frequently needed in behavioral research, but it is certainly not the only possibility. An experimental design may call for more than one independent variable—for example, several teaching methods. It may also assess the effects of those variables in

Box 10-1 Calculation of a Chi-Square [see Table 10-5 and Formula (10-1)]

$$\chi^2 = \Sigma \frac{(f_o - f_e)^2}{f_e} \tag{1}$$

$$= \frac{(50 - 60)^2}{60} + \frac{(50 - 40)^2}{40} + \frac{(70 - 60)^2}{60} + \frac{(30 - 40)^2}{40} \tag{2}$$

$$= \frac{(-10)^2}{60} + \frac{10^2}{40} + \frac{10^2}{60} + \frac{(-10)^2}{40} \tag{3}$$

$$= \frac{100}{60} + \frac{100}{40} + \frac{100}{60} + \frac{100}{40} \tag{4}$$

$$= 1.67 + 2.5 + 1.67 + 2.5 \tag{5}$$

$$= 8.34 \ (p \leq .01) \tag{6}$$

Row 1: This is Formula (10-1).

Row 2: Numerator of each of the four ratios in this row shows the difference between the observed frequency and the expected. (See the four cells of Table 10-5.)

Row 3: Difference from row 2 is squared and divided by the expected frequency, as specified in Formula (10-1).

Row 4: Same as row 3, but notice that there are no longer any negative numbers in the equation. That is appropriate, because *any* deviation of an observed frequency (f_o) from the null hypothesis prediction (f_e) *increases* our confidence that the "votes for Mr. Sex" is related to "sex of voters." The null hypothesis is that the two are *not* related.

Row 5: Each ratio is in decimal form.

Row 6: The sum of the four ratios is the chi-square. (This one is significant at the .01 level.)

various combinations—for example, combinations of methods and teachers. Tests based on the z and t ratios described in Chapter 9 cannot be used to assess significance in either of those situations. However, there is a technique that is appropriate to both of them. It is called *analysis of variance* (abbreviated ANOVA).

One-Way Analysis of Variance

In this section, we shall be concerned only with the design mentioned above in which there is one treatment variable but more than two categories of that variable. Our example will be a simple extension of the experiment we used to illustrate the application of the z (or t) ratio. (See pages 112ff and 117ff.) There we had two groups, which we called control and experimental; here, we shall have six groups, to which we shall refer simply as I, II, III, IV, V, and VI. There we were comparing a new method of teaching French grammar with a traditional method; here we have six different methods, one or more of which may be traditional. Let us say that the control and experimental groups of the earlier experiment are groups I and IV of this one.

Eventually, we shall want to know which of the six methods is (are) the most effective. We could proceed by comparing all possible combinations of groups, but that would be tedious since it would require, in this problem, 15 such tests to check all the possibilities. What we need is some kind of survey test that will tell us whether there is any significant difference *anywhere* in an array of categories. If it tells us no, there will be no point in searching further.

There are other reasons for using an overall test of significance that are more important in the long run than the saving of labor. First, any statistic based on *all* the evidence will be more stable (see "Effect of *n* on Standard Error," pages 103ff) than one based on only part of it, as would be the case if any two of the six methods were compared. Second, there are so many comparisons that *some will be significant by chance*. If there were a hundred such comparisons, five probably would show significance at the .05 level and one at the .01 level even if there were no real differences at all. So whenever we are dealing with several categories of the treatment variable, we need an overall test of significance.

Such a test does exist. It is called the *F test* or *F ratio*. *F* is a ratio of two variances. In Chapter 4, variance was defined as the square of the standard deviation:

$$S^2 = \frac{\Sigma x^2}{n}$$

Then in Chapter 7 (pages 91–93) you learned that

1. the sample statistic S (and hence S^2) is frequently only a means to an end,
2. the end to be approached is the population parameter σ, and

3. when only a sample is available, the parameter can be approximated by a statistic s, the standard deviation of the population as estimated from sample data (or more simply, the "estimated standard deviation of the population.")

Now, since the variance of any distribution is the square of its standard deviation, our best estimate of the variance of a population is

$$\text{estimated population variance} = s^2 = \frac{\Sigma x^2_{\text{sample}}}{n - 1} \qquad (10\text{-}2)$$

where s^2 is the square of the estimated standard deviation of the population, $\Sigma x^2_{\text{sample}}$ is the sum of the squared deviation scores in a sample, and $n - 1$ is degrees of freedom in this calculation.

The F test is a ratio between two variance estimates. But before we analyze the logic of the F test, let's take just a moment to consider in very general terms what it is intended to accomplish. We have six groups. The mean of each group differs by some amount from that of every other group. The question is "Are those *significant* differences?" (or more precisely, "Is there at least one significant difference among them?"). We want to know whether the observed variability of the means is greater than could be expected by chance. Look at Figures 10-1 and 10-2. Which of these diagrams suggests the more reliable (significant) differences among means?

Of course the smaller the variability of individual scores within each group, the more confident we can be that we are really dealing with different groups— or, more precisely, with samples drawn from different populations. It is relatively difficult to imagine that the six groups in Figure 10-1 are random samples taken *from a single population.* That is the null hypothesis in analysis of variance—that all the groups being compared are samples taken from the same population. The null hypothesis is intuitively more credible in Figure 10-2 than it is in Figure 10-1. But we must not rely on intuition; we need a quantitative test of the null hypothesis. For analysis of variance, that test is in a *ratio* known as F.

One important application of the F ratio is to an overall test of significance. As always in a significance test, the null hypothesis states that all samples are random samples from a single population; as always, we shall attempt to

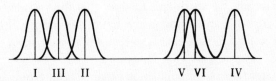

I　　III　　II　　　　　　V　VI　　　IV

Figure 10-1 Groups I through VI with small variability within groups.

I   III   II   V VI   IV

Figure 10-2   The same variability of group means as in Figure 10-1 but with more variability *within* the groups.

disprove that statement. Again, as in the z and t tests, it is a *ratio* that is being evaluated. But this time the critical ratio is not of a difference to its standard error; this time it is a ratio of two estimates of the population variance—one estimate from differences among the means of the categories being studied, the other from differences among individual scores *within* categories:

$$F = \frac{s_b^2}{s_w^2} \qquad (10\text{-}3)$$

where F is a universally recognized symbol for the ratio at the right of the equation sign, s_b^2 is the population variance estimated from observed variability among ("between") the means of the groups, and s_w^2 is the population variance estimated from observed variability within the groups.[3]

You will remember that back in Chapter 8 (especially pages 97–99) we used the variability of individual scores to estimate the variability of means. Well, it can work the other way, too; we can use the variability of obtained means to estimate that of the individual scores in a population. The numerator in the F ratio is just such an estimate: the population variance estimated from the variability of *means* of the groups being studied. The denominator is the population variance estimated from the variability of *individual scores* within those groups. So we have in this ratio two estimates of population variance. You could say that the numerator is an estimate of population variance with the effect of the treatment variable (differences among the means) included, while the denominator excludes that effect. If the two estimates are the same, there is no such effect. The null hypothesis states that they *are* the same—that the ratio is 1.00.

If we are to demonstrate that some real difference does exist among the effects of our six teaching methods, we shall have to show that the ratio is greater than 1.00. Specifically, we shall have to show that the variance estimated from group means is greater than the variance estimated from individual scores within the groups. Understanding why that is so requires an analysis of the logic behind the F test.

Every estimate we make of population variance is an estimate of the variability of individual scores around a true mean. The closest thing we have to a true mean here is the *grand mean*. Every individual deviation from the grand mean can be thought of as two components: (1) the deviation of the individual score from its group mean and (2) the deviation of that group mean from the grand mean. Now, if we could somehow eliminate the means component, we should have an estimate of *what the population variance would have been* if there had been *no differences among the group means*. If all of the group means are alike, then each is identical to the grand mean and the deviation of any score from its group mean is identical to its deviation from the grand mean. From such deviations we can estimate population variance. The result is the *denominator of the F ratio*.

If you will compare Figures 10-2 and 10-3, you will see immediately that the total variability of the six groups is smaller when the variability of their means is eliminated. (You see only *one* mean in Figure 10-3, because all six are in the same place.) That is why the denominator of the F ratio is smaller than the numerator if there are any differences at all among the group means.

Those group means do vary almost always; the question is whether they vary enough to be significant in the sense that was developed in Chapter 9. To find out, we compute a ratio. To provide the *denominator* of that ratio, we discover how much *individuals* deviate from their own *group* means, as in Figure 10-3; whereas the *numerator* also takes into account the deviations of those *group* means from the *grand* mean.

The null hypothesis is that there is no variance among the means: Hence it predicts that an estimate of the population variance that *includes* that variance (the numerator of the F ratio) will be the same as one that *excludes* it (the denominator); if there are no differences among the means, the population variance estimate that takes those differences into account will be no larger than the one that does not, and the F ratio will be 1.00. Ratios larger than 1.00 can be evaluated by tables that give the probability that any given F occurs entirely by chance. That, of course, is the *significance level* of the ratio, and it is interpreted

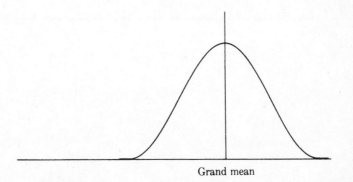

Grand mean

Figure 10-3   The six distributions in Figure 10-2 superimposed so that differences between means are eliminated.

in the same way as one derived from a z or a t. We reject the null hypothesis if the probability of its occurrence by chance is sufficiently low; usually $p \leq .05$ will suffice.

Box 10-2 was designed to illustrate the calculation, from Formula (10-3), of an F ratio for the six groups depicted in Figure 10-1. The illustration is not very

---

Box 10-2   **Calculation of one-way analysis of variance**
             (see Figure 10-1)

Data table

		Groups					
	I	III	II	V	VI	IV	
Individuals within groups	9	10	12	15	17	17	
	11	12	13	16	15	19	
	10	11	11	14	16	18	
Σ	30	33	36	45	48	54	
$\overline{X}$	10	11	12	15	16	18	Grand $\overline{X} = 13.7$

Source table

Source	df	SS	MS	F
Between	5	148	29.6	29.6
Within	12	12	1	
Total	17	160		

$$F = \frac{MS_b}{MS_w}$$

$$= \frac{29.6}{1}$$

$$= 29.6 \qquad (p \leq .01)$$

The data table organizes the data of the experiment in the same way that the subjects were organized into 6 groups of 3 individuals. Notice that:

realistic, however, because in order to make it as easy as possible for you to follow all the manipulations, it was necessary to make the n of each group very small. (A classroom group is much more likely to include 30 students than the 3 that make up each group in Box 10-2.) If you could see the calculations for six groups of 30, however, you would be grateful that these groups are so small.

1. individuals vary around their group mean (*within-group variance*), and
2. group means vary around the grand mean (*between-groups variance*).

The source table is the analysis of variance table that you are most likely to see in a research report. Two of the terms in this table, "SS" and "MS," are new. The concepts to which they refer are familiar, however:

SS = Sum of squares, i.e., sum of the squared deviations from the mean = Σx^2

MS = mean square, i.e., mean of the squared deviations from the mean = $\Sigma x^2/df$, which is the *variance* (s^2; see page 39)

The variance term (MS) in the "between" row of the table is of group means around the grand mean; the variance term (MS) in the "within" row is of individuals around their group means. (The "within" row is often labeled "error.") The F ratio is thus

$$\frac{s_b^2}{s_w^2} = \frac{\mathrm{MS}_b}{\mathrm{MS}_w}$$

The term df refers to *degrees of freedom*, which is the denominator of the formula for the estimated population variance, whether the estimate is based on between- or within-group deviations:

Between df = number of group means minus the 1 degree of freedom lost by computing the grand mean: $6 - 1 = 5$

Within df = number of individuals in one group minus the degree of freedom lost by computing that group mean multiplied by the number of groups: $6(3 - 1) = 12$

Don't forget that a significant F reveals only that there is at least one difference between groups somewhere in the table of data. (See the discussion in text on page 132.)

One feature of Box 10-2 that may puzzle you is that the formula for F, instead of the familiar

$$F = \frac{s_b^2}{s_w^2}$$

appears there as

$$F = \frac{MS_b}{MS_w}$$

Be assured that the two expressions are really equivalent. MS is a symbol for *mean square*. Now recall what is under the square root radical in the formula for the estimated standard deviation of the population:

$$s = \sqrt{\frac{\Sigma x_{sample}^2}{n - 1}}$$

Remember, too, that what is under that radical is the estimated variance of the population:

$$s^2 = \frac{\Sigma x_{sample}^2}{n - 1} = \text{estimated population variance}$$

Now notice that s^2 is a *mean of the squares* of individual deviations from the population mean. When calculating the F ratio, statisticians nearly always use MS (read "mean square") instead of s^2 to refer to variance; thus,

$$F = \frac{s_b^2}{s_w^2} = \frac{MS_b}{MS_w}$$

where s_b^2/s_w^2 is the familiar F ratio of two variance estimates [Formula (10-3)]. Since the mean of the squared deviations (MS) *is* the variance, the ratio of mean squares is a ratio of variances—that is, it is F.

After the F Test

When an F test turns out to be significant, we know, with some specified degree of confidence, that there is a real difference somewhere among our means. But if, as in our example, there are more than two categories ("groups") within the variable being examined, we don't know *where* that difference is.

The most obvious approach to this problem is to perform a t test on each difference, beginning with the largest and continuing until one test fails to achieve significance. That is not acceptable, however, for some of the same reasons that prevented us from using t in the first place. Take another look at the third paragraph of the preceding section; the second reason cited there is the most cogent one for not using t after F: "There are so many comparisons that some will be significant by chance." Some statisticians will approve the use of t in analysis of variance designs, either before or after analysis of variance, but only if the particular comparisons are selected on a rational basis *before the data are collected*. It is the same requirement mentioned earlier in relation to the use of a one-tail test of significance (page 120); the reason for it is essentially the same, and the same controversy obtains.

Other tests have been devised for use in the post-F situation. All are attempts to disprove the null hypothesis, and all have been made more difficult to pass than t in order to compensate for the number of comparisons and the concomitant increase in the probability that some will show significant differences by chance.

More Complex Designs

We have examined the logic of the F test using a one-way analysis of variance as an example. Others are possible: two-way, three-way, four-way, and so forth. My first impulse was to illustrate a two-way analysis by expanding our one-way example into a methods-by-teachers design; each method would have been used by five teachers, making 6 methods $\times$ 5 teachers = 30 treatment conditions. But that would have gotten us into problems inappropriate to a discussion of this kind. For example, the performance of a teacher using any one method would probably be affected by whatever experience he or she had had with the other methods; that would have taken us into the problem of counterbalancing the design. Also, the larger number of cells in such a matrix would have made your comprehension of interaction effects more difficult than I believe is necessary. So let's leave that one, with the passing comment that once the design problems have been solved and the treatments applied, data from such an experiment can be evaluated by analysis of variance techniques.

What we want now is the simplest design that can be used to illustrate the basic principles of two-way analysis of variance. Such a design is called a 2 $\times$ 2 *factorial design*. Also, we'll select treatment variables that do not require a counterbalanced design. As in the preceding illustration, calculations will be extremely simple.

The dependent variable in this experiment (the one affected by the treatment) is *persistence in the face of failure*—specifically, the amount of time spent on an insoluble problem. The two treatment variables are *stress* and *self-confidence*. The design is shown in Table 10-6 as a 2 $\times$ 2 matrix in which there are four cells, each of which represents one treatment condition—(1) low stress with low confidence, (2) low stress with high confidence, (3) high stress with

TABLE 10-6   A 2 × 2 factorial design with index numbers
of subgroups of subjects as cell entries

		Confidence		
		Low	High	
Stress	High	3	4	(Group 3-4)
	Low	1	2	(Group 1-2)
		(Group 1-3)	(Group 2-4)	

low confidence, and (4) high stress with high confidence. (Find the corresponding index numbers in the table). There are 25 subjects in each treatment condition; each subgroup is a random sample of a population of 10-year-old American males, and the question to be answered is whether they will *remain* random samples of a single population, with respect to persistence, at the end of the experiment. The null hypothesis says they will.

Several hours before the experiment begins, all subjects receive a painful electric shock "accidentally" while playing with some laboratory equipment. (As will become apparent later, it is important that they know what it is like to be shocked.) Then, just a few minutes before the persistence task is introduced, they all take a short paper-and-pencil test. The test is also the same for all subjects; however, half of them (group 2-4) are told that their performances were successful and the other half (group 1-3) that their performances were not. We are manipulating their *self-confidence*.

Immediately after that experience, they are all introduced to the persistence task. The problem is insoluble, but the subjects don't know that. They are all told that they should do the best they can, but that they may leave at any time they wish. Then half of them (group 3-4) are told that if they fail the test they will receive several shocks of the kind they had experienced earlier, whereas the other half (group 1-2) are told nothing of the shocks; thus, we are manipulating *stress* as a second independent variable.

Now, analysis of variance offers three advantages over the z or t type of significance test: (1) It can compare the effects of more than two categories of a treatment variable[4]; (2) it can compare the simultaneous but separate effects of two or more variables; and (3) it can assess the interaction effects of two or more variables. The first of those advantages was illustrated in the preceding section. The second is attained by computing an F ratio for each treatment variable. In the present case, we would do one F test for the "stress" *main effect* (group 1-2 versus group 3-4) and another for the "confidence" main effect (group 1-3 versus

TABLE 10-7 Group means (minutes at task) in two-way analysis:
outcome 1

		Confidence		
		Low	High	
Stress	High	³ 20	⁴ 30	50
	Low	¹ 10	² 20	30
		30	50	

group 2-4). The most interesting of the three features of analysis of variance, but also the most difficult to understand, is the last named: its ability to identify interaction effects.[5]

The best way to explain interaction is to cite an example, so let's return to our experiment. Table 10-7 is like Table 10-6 except that the index number of each subgroup has been moved to the upper left-hand corner of that subgroup's cell, and the number in the middle of the cell is the subgroup's mean score (time spent before quitting). The data in the table show what look like two substantial main effects. (Whether they are significant depends also on the amount of variance *within* the two groups that are being compared in each case; let us assume a sufficiently small amount.) Group 3-4 (50 minutes) is more persistent than group 1-2 (30 minutes), and group 2-4 (50 minutes) is more persistent than group 1-3 (30 minutes). High stress produces more persistence than low stress, and high confidence produces more persistence than low confidence. But there is no interaction.

By contrast, imagine that the outcome of the experiment is as depicted in Table 10-8. There is one main effect (high versus low confidence), and there is interaction because increasing stress (from low to high) has a different

TABLE 10-8 Group means (minutes at task) in two-way analysis:
outcome 2

		Confidence		
		Low	High	
Stress	High	³ 10	⁴ 30	40
	Low	¹ 20	² 20	40
		30	50	

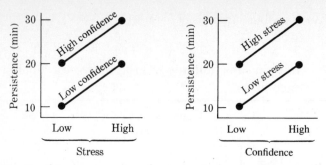

Figure 10-4 Graphic representation of outcome 1, showing two main effects and no interaction.

Box 10-3 Sample Calculation of a Two-Way Analysis of Variance, Outcome 1 (see Table 10-7 and Figure 10-4)

Data table

		Confidence	
		Low	High
Stress	High	3 17 23 21 $\underline{19}$ $\Sigma = 80$ $N = 4$ $\bar{X} = 20$	4 29 27 33 $\underline{31}$ $\Sigma = 120$ $N = 4$ $\bar{X} = 30$
	Low	1 9 7 13 $\underline{11}$ $\Sigma = 40$ $N = 4$ $\bar{X} = 10$	2 19 17 23 $\underline{21}$ $\Sigma = 80$ $N = 4$ $\bar{X} = 20$

Source table

Source	df	SS	MS	F
Stress	1	400	400	59.97
Experience	1	400	400	59.97
Stress × experience	1	0	0	0.00
Within	12	80	6.67	
Total	15	880		

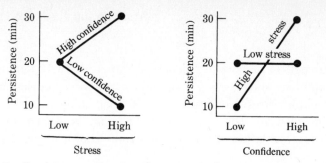

Figure 10-5 Graphic representation of outcome 2, showing interaction and one main effect.

$$F = \frac{MS_b}{MS_w}$$

$$F_S = \frac{MS_{b_S}}{MS_w} = \frac{400}{6.67} = 59.97 \qquad (p \le .01)$$

$$F_C = \frac{MS_{b_C}}{MS_w} = \frac{400}{6.67} = 59.97 \qquad (p \le .01)$$

$$F_{S \times C} = \frac{MS_{S \times C}}{MS_w} = \frac{0}{6.67} = 0$$

where b_S indicates "between groups that differ in amount of stress," b_C indicates "between groups that differ in self-confidence," and $S \times C$ indicates "interaction of stress with self-confidence." You already know the meaning of MS_w.

The data table displays the raw scores of all of the subjects, how the experiment was organized, and the means of all of the subgroups. The dependent variable throughout is "persistence," as measured by time on task.

The first two sources listed in the source table are between-groups variation. The persistence of the subjects may differ as a function of the amount of stress to which they were exposed or as a function of their recent experience of success or failure. There is also the possibility that the factors of stress and confidence interact. That possibility is represented as "stress × confidence." The final source, "within," comprises the deviations of individuals from their various group means. (The "within" row is often labeled "error" or "residual.")

The sums of squares (SS) and the mean squares (MS) are calculated as they were in the one-way analysis of variance illustrated in Box 10-2. In this case, however, there are three calculated Fs rather than just one, with one F specifying the significance of the variation caused by each of the independent variables and a third concerning their interaction. It should be apparent that in this experiment both stress and confidence were associated with significant variation between groups but that there was no interaction between them.

effect on subjects in the low-confidence condition (down 10 minutes) than it does on confident subjects (up 10). Similarly, confidence has a different effect in the high-stress condition (up 20) than it does in the low-stress condition (no change). The interaction effect may be seen more clearly in the graphic representation of the two outcomes shown in Figures 10-4 and 10-5.

Very different patterns, aren't they? In outcome 1, there is a very simple main effect of stress shown in the left diagram and of confidence on the right. Outcome 2, however, shows that the effect of high stress—as opposed to low— is to *increase* persistence in subjects who have recently experienced success and to *decrease* it in those recently subjected to failure. It also shows that the effect of the experience of success (high-confidence condition) is to increase persistence dramatically in the high-stress group but to make no change at all in the low-stress group.[6]

We might be tempted to speculate about possible explanations for those results. However, this is a treatise on statistics, not psychology; also, these data were not derived from any real experiment. The main point is that interactions do occur, and that when they do, they can be detected by analysis of variance.

The calculation of three F ratios (stress main effect, confidence main effect, and stress-by-confidence interaction) for outcome 1 is illustrated in Box 10-3. The sizes of the groups here are perhaps a little more realistic than those in Box 10-2, but they are still very small. That should help you to follow the calculations.

Summary

Chapter 9 showed how a difference between two groups of subjects can be evaluated in terms of the probability of its occurrence by chance (sampling error). Chapter 10 has extended that technique to situations in which the data are in the form of *frequencies,* to those in which *several* groups differ on a given factor (treatment variable), and even to some in which several factors must be evaluated simultaneously.

Frequency data are analyzed via the χ^2 technique. Multiple groups are addressed via *one-way analysis of variance* and the calculation of an F ratio. An example is also given of a more complex design to which analysis of variance might be applied—a two-way analysis in a 2×2 *factorial design*. A distinction is drawn between *main effects* and *interaction effects*. Because F functions initially as a kind of survey test, the follow-up problem—the identification of the precise source of these effects when F has proven significant—is also discussed.

Sample Applications

EDUCATION

1. In an attempt to decrease the number of high school dropouts, a school system develops a vocational training program for students who are at high risk for becoming dropouts. During the first several years of the program, the school staff wants to determine whether the program is effective in reducing the number of students who drop out, and you are asked to help. You select at random the 50 students that the program can accommodate from all students who apply and are at risk. Then you keep track of which of those applicants eventually graduate from high school. What statistic will help you decide whether the special program is effective?

2. You are a school psychologist. You have developed a program to help students systematically think about and plan solutions to social problems with which they may be faced in school, in dealing with friends, and in job situations. To try it out, you train half of the counselors in the school system to use the program. Teachers first identify students who are having difficulty solving social problems. Then one-third of those students are assigned randomly to trained counselors and another third to untrained counselors. The remaining third are provided no counseling at all. At the end of the semester, all students are tested to determine their ability to solve a number of social problems. How will you interpret the resulting data?

3. You are an educational psychologist. You and a group of your colleagues have developed three different educational programs to enhance the abstract reasoning ability of sixth- and seventh-grade students. With the cooperation of a large urban school district, 200 sixth- and seventh-grade classrooms are assigned to the four treatment programs (one of them a control condition), 50 classrooms per program. All students are tested on abstract reasoning at the beginning and end of the year. How could the resulting data be analyzed and interpreted?

POLITICAL SCIENCE

1. You are interpreting the results of an opinion poll. In a random sample of 100 individuals, you find that of the 40 Republican respondents a total of 30 favor a proposal to cut the capital gains tax, and that of the 60 Democratic respondents 20 support it. How can you calculate the probability that this association is due to chance?

2. Once again you are studying the incidence of military coups in Latin America. This time you want to know whether there are significant differences among those nations in the number of coups that occur, and you have reason to believe that the type of legitimacy upon which the regime rests (i.e., traditional, legal-rational, or charismatic) is the most important factor in accounting for the incidence of coups. In this situation you have a treatment variable (type of regime legitimacy) and a dependent variable (coup frequency). How can you ascertain the probability that types of regimes differ with respect to coups?

3. One of the most frequently studied questions in the field of international relations pertains to the relationship between domestic politics and foreign policy behavior. You are in that field and you want to test the idea that a nation's form of government (democratic, authoritarian, or totalitarian) and type of leadership (unitary, collective, or fragmented) affect the number of aggressive actions it initiates (the dependent variable). What kind of analysis is appropriate?

PSYCHOLOGY

1. You are the director of a clinic dealing solely with phobias (unusual fears) of children. Over a two-month period, 100 children are referred to your clinic for the treatment of agoraphobia (fear of large open spaces). After hypnosis therapy, you find that 55 of the 100 children are able to walk in a large open field and report little or no fear. These are fairly impressive results, but are they statistically significant? How can you tell?

2. You are a child clinical psychologist interested in which of three approaches is most effective in controlling the pain experienced by children in a hospital burn unit. You assign, at random, equal numbers of children to (1) a mental distraction condition, where the participants attempt to keep their minds off the pain by doing mental arithmetic; (2) a self-reward condition, where participants give themselves positive self-statements (e.g., "I'm a brave person") for enduring pain; and (3) imagination exercises, where the participants are taught to imagine situations in which they experience pleasure and satisfaction. How might you go about analyzing the results?

3. You are a psychotherapist at a child guidance center. You are interested in the possible interaction between personality factors and the effectiveness of psychotherapy. You assign each of 20 introverted (shy and socially withdrawn) children to either individual or group therapy. Next, you assign 20 extroverted (socially outgoing) children to the same conditions. Your prediction is that the introverted children, because of their social fears and shyness, will

benefit most from individual therapy and that the extroverts will work best in a group setting. Presuming that you have a generally accepted criterion of effectiveness of therapy, how might you analyze the data from this study?

SOCIAL WORK

1. You are a social worker in a small rural community. You conclude that the existing manner of dealing with juvenile offenders through the courts is of limited efficiency and effectiveness. The magistrate is present only one and one-half days a week, and the court calendar is always too full to handle the number of youth being petitioned into court.

    You develop an alternative to this process by convincing local citizens and the judge to establish a juvenile review board. The board, composed of community people, will decide cases and use measures such as community work service, restitution, and active parental involvement instead of the jail time and probation that have been the remedies usually prescribed by the courts. Part of the agreement in establishing the alternative program is an evaluation of its effectiveness, at the end of one year, in comparison with the effectiveness of the courts' traditional procedures. Low rate of recidivism is selected as the major indicator of success, and youth will be assigned to the alternative program or to the courts on a random basis, with violent offenders excluded from both groups.

    When the one-year period is over and all the data are in, how do you evaluate the results?

2. In your child guidance clinic there is disagreement concerning how to treat children who display hyperactive behavior. One social worker with a psychoanalytic background favors play therapy with two sessions a week for a minimum of one and one-half years. The agency psychologist favors a behavioral model and the use of operant conditioning over a much shorter period. A consulting psychiatrist views hyperactivity as the result of immature cortical development, and she recommends the drug methylphenidate to stimulate the cortex.

    Since hyperactivity is so frequent and so troublesome, you recognize it as a major problem. You decide to conduct a study to see which is indeed the most effective of the three treatment methods; you also decide to add a fourth group of children who will receive no treatment. (The clinic has a waiting list.) Children will be assigned randomly to those four groups. The social worker, the psychologist, and the psychiatrist all agree on a single test score as the criterion of effectiveness. Once you have obtained all your data, how do you evaluate it?

3. As a social worker in a foster care agency, you are interested in how the age of children and youths affects their adjustment to either a foster family home or a staffed community group home.

    The independent variables are age and type of foster care arrangement. Age is defined by two categories: "children" (5 though 12 years) and "youths" (13 through 18 years). You obtain a standardized instrument to measure the dependent variable, adjustment to foster care.

    Twenty children are randomly assigned to either a foster family home (10 children) or a group home (10 children), and 20 youths are randomly assigned to either a family home (10 youths) or a group home (10 youths).

    Are there any reliable differences in adjustment among those four groups? If so, can they be traced to age differences, the types of foster care, or both? Is there an interaction between the two? Those are the questions that your design is supposed to address; how do you treat the data statistically to help obtain the answers?

SOCIOLOGY

1. You are still a consultant to the family planning group described in the sociology exercise for Chapter 9. This time you are asked about the antecedents of different attitudes toward abortion. You cannot answer such a broad question all at once, but one of your many hypotheses is that a person's belief about when human life begins influences his attitude toward abortion. Some questionnaire data are available from another study, and you are able to find two questions pertinent to your hypotheses: (1) "Do you believe that human life begins before or after the 90th day following conception?" and (2) "Do you approve of abortion on demand?" The respondents' answers to these questions are your data. What do you do with them?

2. That family planning agency is intrigued by your answer to its earlier question about the effect of religious belief on family size (see the sociology exercise in Chapter 9). The agency wants to broaden the investigation by comparing the sizes of Catholic, Mormon, and Mennonite families with each other and with that of the general population. You gather the data and find that there are differences among the means. How can you tell whether these differences are significant?

3. You are asked to identify some of the critical variables that inhibit or enhance communication of sex information to children by their parents. "Amount of sex information communicated" is therefore the dependent variable in your study. You construct an instrument that yields a sex information score and administer it to parents of 10-year-old children. (The score represents the amount of information that parents believe they would give to

their 10-year-old if the child were to ask "Where do babies come from?") You hypothesize that the amount of information contained in the parents' answers to this question is (1) inhibited by church involvement and by lack of education and (2) enhanced by little or no church involvement and high education. You define "church involvement" as attendance 30 times a year or more, "nonattendance" as attendance 3 times a year or less. Two or more years of college defines the "high education" group, and less than a year of college identifies the "low education" group. How will you analyze the data that emerge from this study?

11

Correlation, Causality, and Effect Size

Correlation is not causality.

Before we can be certain that we have found an instance of *causality*, we must observe a reliable relationship between the purported causative agent and the event being caused. If every time you push a switch in your bedroom to the up position a light goes on, you come to believe that the latter event is caused by the former, even in the absence of wiring diagrams, electrical theory, and so forth. If your dog becomes wildly aggressive whenever the garbage man approaches, you say that the man's approach is causing the dog to be upset. If in studying a large sample of schoolchildren you find that those with larger vocabularies have higher IQs, you assume that the vocabulary is a cause of the intelligence.

But should you? Is it legitimate for you to assume that when two events occur together one is caused by the other? Consider the following case: What size of coefficient would emerge if we were to compute a correlation between the revolutions per minute of the left and right front wheels of your car (Figure 11-1)? The correlation would be pretty high, wouldn't it? If the car were on a curving road and turns in one direction were predominant throughout any given minute, the two scores would be different but correlated for that minute, and with varying speeds on a straight road, the correlation would be almost perfect.

But does the speed of one wheel *cause* that of the other? Only to the extent that each is a functioning part of a total system in which the two events occur. It would be more accurate to say that the functioning system—the car in motion—causes the revolutions of both wheels.

I don't suppose anyone will ever bother to find the correlation between the two front wheels of an automobile, but the same principles apply to many situations in which correlation coefficients *are* commonly computed. For example, if we were to study a large sample of elementary school children, we might find a

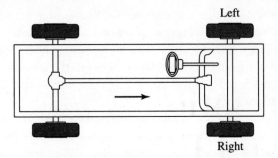

Figure 11-1   Top view of an automobile chassis.

tendency for those children who display extensive vocabularies to be superior students in arithmetic. We might be tempted to conclude that a good vocabulary causes competence in arithmetic. But then we find that chronological age correlates with both variables. Might it not be better to say that achievements in both vocabulary and arithmetic are caused by (are parts of the functioning system of) the child's developing intelligence?

The caution that I am advocating here has important practical implications; without it we might easily be persuaded, for example, to set up elaborate vocabulary-building programs for children who show weakness in quantitative thinking. Some wag once noted that there is a substantial correlation between the intelligence of boys (as indicated by their mental ages) and the length of their trousers. He suggested that a relatively inexpensive way to increase the intelligence of boys would be to increase the length of their trousers!

So although correlation is a *necessary* feature of a causal relation, it is not *sufficient* to demonstrate causality.[1] Whether the relation is interpreted as a causal one should depend not just on the correlation of two variables but also on some rational link between them—on the extent to which the relationship makes sense within some sort of conceptual framework (within a wiring diagram, for example, or a sociological theory) or, ideally, on one of these conceptual structures plus the elimination of alternative possibilities, as elaborated in the following section.

Correlational versus Experimental Studies

People who design research studies usually distinguish between correlational and experimental designs. Most *correlational* studies are concerned directly with relationships among variables that occur naturally—that is, without the intervention of the investigator. An example might be the relationship of poverty to intelligence. If there is a unique entity called "poverty" that can be measured, and if there is another unique entity called "intelligence" that also can be measured,* then a researcher can discover the relationship between poverty and intelligence and report that relationship numerically—probably with a Pearson r. If the r is significant, the researcher can make a direct inference of *relationship* but no inference of causality in the relation between the variables. (Does poverty cause low IQ, does low IQ cause poverty, or does yet another causal structure underlie the relation?)

It is this conventional use of correlational information that was described in Chapter 6. However, unconventional uses have become more common in recent years—uses that may eventually redefine what constitutes conventional use. This chapter describes some of those new ways to interpret correlations.

Traditionally, *experimental* investigations are designed to identify *causes*. If it were possible for an investigator to *manipulate* poverty, it could be determined whether its relation to intelligence is causal. In this design the researcher would manipulate certain variables instead of merely observing them. In psychology, the simplest such case would have all relevant treatment variables held constant except one. That one would be manipulated by the experimenter, who could then make a direct inference that any change in the behavior of the subjects (in this case, their performance on an IQ test) is caused by a change in the manipulated variable (in this case, their economic well-being). The manipulated variable is called *independent*,† variables that are held constant are *controlled,* and the behavior of the subjects is the *dependent* variable (because their behavior depends on what happens in the independent variable).

I have just said that variables that are held constant in an experiment are known as controlled variables. And so they are, but a so-called controlled variable can also vary at random (be entirely *un*controlled); the important thing is that its variance be unrelated to that of the independent variable. In the example, the investigator would manipulate the environments of the subjects; if the children who were low on a scale of economic level later turned out to be low on a scale of intelligence, then it would be tempting to infer that poverty causes low intelligence. But if infants placed in poverty were *genetically* inferior to those assigned to other environments, no such inferences could be made because there would be no way of knowing whether the genetic difference or the environmental difference had caused the observed difference in intelligence. A sample that fails to take some potential influence (in this case that influence is genetic) into account is called a *biased* sample.

To make legitimate the inference that poverty causes low intelligence, the experimenter would have to make sure that genetic variance (a controlled variable in this experiment) is unrelated to variance in economic status (the

---

*To simplify the discussion and keep it focused on statistics, we shall assume the validity of both of these propositions.

†In many applications, the term *treatment* can be used in place of *independent*.

independent variable). If genetic potential varies at random (e.g., high potential is just as likely to be found in the poor environment as in the good one), then a difference of intelligence in favor of the good environment would imply that good environment is a *cause* of high intelligence. The purpose of randomizing a "controlled" variable is the same as that of holding it constant: to prevent any systematic relation between that variable and the one that is being manipulated (the independent variable).

The purpose of all this planning and manipulation is to exclude other possible explanations of whatever change is observed in the dependent variable. If all but one of the potential independent variables are controlled, then that one must be responsible for any observed change in the dependent variable. That is an ideal to be approximated as closely as possible; an explanation is said to have *internal validity* to the extent that alternative explanations can be excluded.

On the other hand, the attempt to achieve internal validity may threaten *external validity*. For example, if a laboratory experiment is so well controlled that its subjects experience it as distinctly different from circumstances outside the lab, its conclusions may not be valid in any other situation. External validity is the extent to which the results of a study can be generalized.

If your main purpose in reading a research report is to understand the *underlying structure* of a particular class of events, then your primary concern will be with *internal* validity. If your intent is to *apply* the results, then you want *external* validity. If you are concerned about both, you may have to accept some kind of trade-off, for internal and external validity are sometimes inversely proportional to each other.

Traditionally, the Pearson r has been used in correlational studies and only in correlational studies. It is, after all, a coefficient of correlation. But if scores on a dependent variable can be shown to be correlated with scores on a manipulated (independent) variable with all other variables held constant (or otherwise dissociated from the independent variable), that correlation *can* be used as evidence of a causal relationship.

It is true, however, that whereas the Pearson r was invented to deal with refined measures of continuous variables, many experimental studies use rather crude measurements (e.g., the categories "low" and "high" or "low," "medium," and "high" for the variable "economic level"); so the issue of *continuity* must be addressed.

Continuous versus Discontinuous Variables and Measurements

The Pearson r is an index of relationship between two continuous variables. (Presumably those variables are derived from parameters that also vary continuously; we shall make that assumption in the discussion that follows.) Many

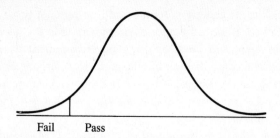

Fail   Pass

Figure 11-2 A possible distribution of point scores in a pass–fail class. The lowest score is less than 50 points, the highest several hundred; but only two grades are reported.

variables are not continuous, however, and even those that *are* continuous may be measured in such a way that their data are not. An example of a dichotomous variable is gender. Virtually all humans are either male or female; there is no continuous dimension underlying the two categories, for the difference between them is *qualitative.* Conversely, although the evaluation of college students in a pass–fail course also produces a report containing just two classes of data (Figure 11-2), each class represents a much larger number of categories that *could* have been reported, and the differences are *quantitative.* If the professor uses a point system to determine who fits into which category (and if there are many students in the course), the distribution of students on scores will be long and nearly continuous. No matter where the professor draws the line between "pass" and "fail," there will be students very close to the line on either side, but there will be only two categories in the professor's report to the registrar. Even a report of letter grades will comprise only five categories (Figure 11-3), although there could be many more (up to a maximum of one for each possible score). Theoretically, if the process of measurement were continued indefinitely, there would be an *infinite* number of scores and potentially an infinite number of categories. Such a variable is continuous.[2]

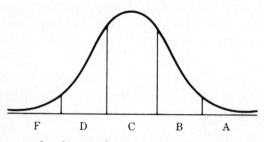

F   D   C   B   A

Figure 11-3 The same distribution of point scores as in Figure 11-2 but with five grades reported instead of two. There are still many different scores within each class interval.

As you have known since Chapter 2, when a continuous variable is represented by data in a large number of categories, those categories are called *class intervals*. When a continuous variable is represented by only a few categories of data, it might be helpful to think of the categories as extremely large class intervals. (Of course, when the variable is by its very nature *not* continuous, as in the gender example cited earlier, the investigator has no choice but to limit the number of categories, none (or neither) of which is a class interval.)

Table 11-1 features four data displays. Each display is itself a table, but it also has graphic qualities. For example, compare the spacing of scores on the independent variable in Table 11-1A with that in Table 11-1B: In A the spacing is nearly continuous, whereas in B it is dichotomous. Be aware of those special relations as you examine the rest of Table 11-1.

Table 11-1 shows four possible ways of organizing measurements taken on two continuous variables from 200 subjects. Independent measures that vary in amount get larger from left to right in each table. Dependent measures that vary in amount get larger from the bottom to the top of the graph. (There is no convention to guide the investigator whose independent or dependent variables are *qualitative*).

In Table 11-1A, both data sets are continuous; in Tables 11-1B and 11-1C, one is continuous, the other dichotomous; in Table 11-1D, both are dichotomous. Table 11-1A carries the most information and Table 11-1D the least, as an examination of the tables will reveal. (Besides examining these tables analytically, you can make an intuitive comparison by holding the page nearly parallel to your line of sight, first looking at it from the bottom, then from the left side of the page. The numbers will be illegible, but the contrast between continuous and discontinuous distributions will be enhanced. The shading in some of the cells identifies those that are featured in the discussion that follows.)

In Table 11-1B, the 2 lowest scores (shaded cell) on the independent variable are treated exactly the same as the 38 that are almost as high as the median, and the 2 highest are treated the same as the 38 that are barely above the median. In Table 11-1C, the same is true of the dependent variable; in Table 11-1D, it is true of both. An r computed for continuous variables from tables in which at least one variable is treated dichotomously (as in Tables 11-1B, 11-1C, and 11-1D) is only an approximation of the r obtained when the data are arranged as in Table 11-1A.

There are two possible reasons for data reports like those in Tables 11-1B, 11-1C, and 11-1D: (1) As in these illustrations, data from a continuous or nearly continuous variable are gathered into a few—here two on each variable—broad categories, or (2) the small number of categories in the data represent an equally small number in the variable being measured. So a dichotomy could occur in a set of data because either (1) administrative concerns are compelling, as in the pass–fail course mentioned earlier, or (2) the variable naturally breaks into two parts, as in the gender example cited previously.[3]

TABLE 11-1   **Four possible displays of data on two continuous**
variables ($r = 1.00$)

A   Frequencies are distributed
across 12 class intervals—a nearly
continuous distribution—on each
of the two variables.

B   There are two distinct groups
on the independent variable, while
the distribution remains nearly
continuous on the dependent.

In general, there are three types of relationship between a variable and the
data that represent it:

Type 1: A continuous variable is represented by a number of data categories
that approaches infinity. (There are, of course, practical limits to the number
that can be used.)

Type 2: A continuous variable is represented by a number of data categories
that approaches one. (The smallest number that can be used is two.)

Type 3: A variable that consists of only a few categories (commonly two) can
be represented by that same number of data categories.

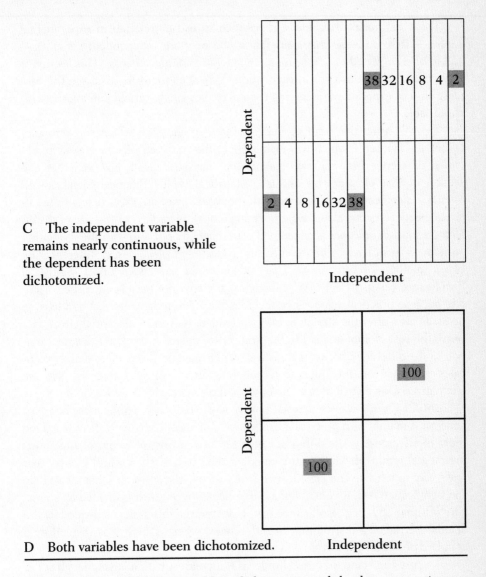

C The independent variable remains nearly continuous, while the dependent has been dichotomized.

D Both variables have been dichotomized.

A fourth type, in which the variable is dichotomous and the data are continuous, is probably never used—not deliberately, anyway.

Correlation as an Index of Causality

The Pearson r was developed for situations in which the relationship among the correlated variables and their data is best described as type 1. However—and right now this is the main point—correlation estimates have since been developed that can be used even when that relationship is better described as type 2 or 3. They are frequently reported as r, each with a subscript to identify the kind of estimate that r represents in that particular case.

Those estimates have made it possible to use correlation in experimental studies—that is, those that manipulate one or more independent variables—and to observe whether one or more dependent variables change. This section is concerned with the use of *r* in such studies. We shall focus on its use in the simplest kind of experiment: assessing the effect of a single variable on another single variable.

You will recall that kind of study as the very one with which we were concerned in Chapter 9, which was mainly about the statistical significance of an obtained difference between two means. You may also recall, however, that one section of that chapter was entitled "Statistical versus Practical Significance." *Statistical significance* is concerned with the *reliability* of an effect (e.g., of the difference between control and experimental groups), regardless of the size of the effect; *practical significance* is concerned with reliability but also with the *size* (either *amount*, as in difference between mean scores, or *frequency*, as in difference between numbers of votes cast). That section could have been called "Statistical Significance versus Effect Size," although that would not have been entirely accurate because practical significance includes the value judgments that still have to be made even after the size of an effect is known. (Given a substantial effect, is it worth the cost of attaining it?) In both titles the "versus" is there to call your attention to the difference between statistical significance on the one hand and practical significance on the other; it is *not* intended to suggest that the two are alternatives. One contributes to the other, and ultimately both are important.

But Chapter 9 was concerned mostly with statistical significance; here our concern is mainly with practical significance and specifically with *effect size*. One index of effect size is the difference between two means (of a control and an experimental group) divided by the standard deviation of the control group. That *looks* very much like the significance test that we discussed in Chapter 9. Like the significance test, it consists of a difference between means divided by a measure of variability. In a significance test, however, the difference is divided by the standard error of a distribution of differences, whereas here it is divided by a standard deviation of individual raw scores.

Because the standard error shrinks as *n* increases, with a large enough set of measures even a very small difference can prove statistically significant (because the divisor is so small). That is as it should be, because the significance test is concerned with the *reliability* of the difference, and reliability does increase with *n*. If you are interested in the *size* of an effect, however, you ask a different question. Instead of "What is the probability of getting a difference this large by chance?" you ask, "Assuming that my obtained difference is reliable, where would the mean score of the treatment group fit into the obtained distribution of the control group?" If it is in one of the tails of that distribution, you call it a large difference; if it is near the mean, you have to admit that it is rather small. In neither case do you know how reliable it is.

There are now several commonly used indices of effect size. The one just discussed is a ratio between an observed difference and a standard deviation, but most of them are variants of the Pearson r. In the remainder of this section, we shall limit our discussion to those variants; we shall emphasize their similarities (indeed, we shall treat them all alike), and we shall symbolize each of them with a simple r.

Table 11-2, in conjunction with Table 11-1, should help you to understand how r can be used to interpret the results of an experimental study. In Table 11-2 two experiments are represented in which the success rate (dependent variable) resulting from treatment by therapeutic intervention (independent variable) was recorded. There are 200 subjects in each study, but notice how differently the subjects are distributed. In the table for study A the success rates for the control and experimental groups are the same. In study B, all of the 100 subjects in the control group are in the low-success-rate cell of the table, while the entire experimental group is in the high-success cell.

You might think of the tables in Table 11-2 as scatterplots (see page 67ff) in which the number of class intervals in each variable has been reduced to two. You may get a better intuitive feel from this report of an experimental study than you did from Table 11-1, because Table 11-2 includes a zero correlation for compari-

TABLE 11-2 **Zero and perfect correlations as indicators of the effect size of a therapeutic intervention**

Study A: No correlation ($r = .00$)

		Treatment	
		Control	Experimental
Success rate	Substantial improvement	50	50
	Little or no improvement	50	50

Study B: Perfect correlation ($r = 1.00$)

		Treatment	
		Control	Experimental
Success rate	Substantial improvement	0	100
	Little or no improvement	100	0

son with a perfect one. Study A of Table 11-2 indicates that the therapeutic intervention had no effect at all; in study B, its effect was maximal. Of course most,
probably all, correlations computed from actual data fall somewhere between
those extremes; but the extremes serve better than actual data to illustrate the
meaning of correlation—in this instance, of a correlation computed from dichotomous data. (Both sets of *data* are clearly dichotomous, but the structure of the *underlying variables* is not so obvious. One variable—"treatment"—is dichotomous,
but the other—"success rate"—is essentially continuous.)

Even though most correlations fall far short of perfection, they can yield
information that is of both scientific and social importance. The psychothera-
peutic intervention suggested above would probably produce an *r* somewhere
between the two extremes illustrated in Table 11-2. Table 11-3 lists 18
independent-to-dependent-variable correlation coefficients and shows the difference between the success rates of the control and experimental groups that
correspond to each *r*. For example, a very modest *r* = .30 indicates a change in
success rate from .35 (35 of the 100 untreated patients improved during the period under study) to .65 (65 of the *treated* ones did); an *r* of .50 corresponds to
an increase from 25 untreated to 75 treated successes; when *r* is .70, the increase is from 15 in the untreated group to 85 in the treated—a difference of
70 percent! Notice that *in every case, the difference in the success rates is identical to r.* Somebody still has to decide whether the payoff is worth the effort, but
there should no longer be a question that there *is* a substantial payoff, even
when *r* is as low as .30.[4]

"Correlation is not causality." That was the opening sentence of this chapter,
and I repeat it here. From the information in this section, however, it should be
clear to you that under certain circumstances a coefficient of correlation *can*
serve as an index of causality. And if Table 11-3 is any indication, in those special circumstances (those that constitute an *experiment*) the coefficient can be
more clearly interpreted than it typically is in more traditional correlational studies.[5] Moreover, Table 11-3 suggests that the practical implications of many experiments may be revealed more clearly if the results are expressed as
correlations rather than differences between means.

This use of correlations is not likely to replace the use of differences between means in experimental studies. But even if it never does, you can use the
information in this section to interpret in concrete terms studies of traditional
correlational design. Table 11-3 shows success rates and corresponding correlation coefficients that might be obtained from an *experiment* (here a study of the
effects of psychotherapy), but it could be used to interpret any correlation. The
interpretation would be of the form "If I were to split the distribution on each of
the two correlated variables into halves, the relations given in the table would
obtain." (But lacking the rigorous controls of an experiment there can be no inference of causality.)

TABLE 11-3   Eighteen levels of correlation as indicators
of the effect size of a therapeutic intervention

		Success rate	
r	Control group	Experimental group	Difference
.00	.50	.50	.00
.02	.49	.51	.02
.04	.48	.52	.04
.06	.47	.53	.06
.08	.46	.54	.08
.10	.45	.55	.10
.12	.44	.56	.12
.16	.42	.58	.16
.20	.40	.60	.20
.24	.38	.62	.24
.30	.35	.65	.30
.40	.30	.70	.40
.50	.25	.75	.50
.60	.20	.80	.60
.70	.15	.85	.70
.80	.10	.90	.80
.90	.05	.95	.90
1.00	.00	1.00	1.00

Source: Adapted from R. Rosenthal, "Assessing the Statistical and Social Importance of the Effects of Psychotherapy," *Journal of Consulting and Clinical Psychology* 51 (1983): 12.

Two examples are given below. As you read them, look at the corresponding table. Each cell in the four-cell table gives the proportions not of the entire sample but of those subjects who are in the upper or lower half of the sample of a given variable.

In Table 11-4, an r of .40 between *sugar intake* and *hyperactivity* means that as compared to children in the lower half of the sugar-intake dimension, 40 percent (.70 − .30) more of those in the upper half are also in the upper half on a scale of hyperactivity. Because this is not an experimental study, you cannot be sure what causes what, but the table can help you judge the importance of the relation.

In Table 11-5, an r of .50 between women's *self-esteem* and the duration of *child-support payments* received means that of the women in the *lower* half of the self-esteem distribution, 50 percent more are in the *upper* half (than in the lower

TABLE 11-4   Median split of both distributions (X and Y) when
$r_{xy} = .40$

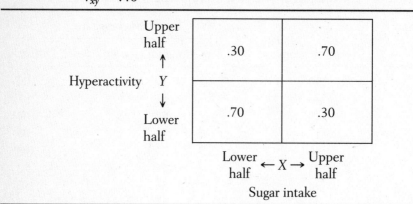

half) in duration of child-support payments. Again, the causal structure is not clear.

In general, the correlation coefficient gives you in a single number all the information you need to construct a 2 × 2 matrix of proportions, four entries in all, as illustrated in Tables 11-4 and 11-5. The proportions in any such table constitute an interpretation of a correlation coefficient.

Correlation is not causality, but with proper precautions it can serve as an index of the strength of a causal relationship. And whether causal or not, a relationship between two variables now can be interpreted in more concrete terms than had previously been possible.

TABLE 11-5   Median split of both distributions (X and Y) when
$r_{xy} = -.50$

Child-support payments   Y

Upper half ↑   .75   .25

Lower half ↓   .25   .75

Lower half ← X → Upper half

Self-esteem

Summary

Traditionally, quantitative research has been done in either a correlational or an experimental mode. *Correlational* designs warrant direct inferences only about the relationships among variables, not about the *nature* of those relationships. *Experimental* studies, on the other hand, are designed so that inferences can be made about the *causality* of the relationships identified by the investigators. In experimental studies, *independent variables* are manipulated while *controlled variables* are held constant (or varied at random) and *dependent variables* are observed. If all relevant variables (other than the independent) are held constant, the changes in the observed variable are said to *depend* on (that is, to be *caused* by) those in the independent variable. In the simplest case, an *experimental group* is exposed to the independent variable while a *control group* is not; both experience *controlled variables* in the same way (either constant or random).

The Pearson *r* is an index of relationship between two continuous variables. Many variables, however, are not continuous, and those that *are* continuous can be represented by data that are not. Indeed, continuous variables nearly always are broken into class intervals for convenience in handling data. If there are many class intervals, the deviation of the data from continuity is trivial; however, if there are only a few class intervals—the lower limit is two—that deviation is more substantial. Even so, useful approximations of the Pearson *r* can now be extracted from noncontinuous data.

An even more recent development is a set of techniques for extracting *effect size* (as distinguished from reliability). Those new techniques include the use of correlation coefficients in interpreting experimental studies. Used in this way, correlation coefficients can be indicators of causal relations: Correlation is not causality, but causality is one kind of relationship between variables. Thus a measure of relationship (the correlation coefficient) can in the right circumstances be a measure of *causal* relationship. But even without knowledge of causality, a correlation between two variables can be important information.

Sample Applications

See pages 87–88, and specify as many plausible causal structures as you can.

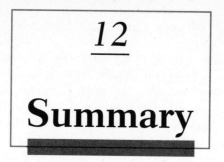

12

Summary

The central objective of this book has been to convey an understanding of the basic concepts and their interrelations—the "big ideas," so to speak—in statistical thinking. In Chapters 9 and 10, for example, the big idea was the logic of testing for the significance of differences, especially in experimental settings; in Chapter 6, it was correlation between variables in settings other than experimental; and in Chapter 11 it was a comparison of correlational with experimental strategies.

The first four chapters established a necessary foundation for everything that was to follow. However, in Chapters 2, 3, and 4 it became apparent that big ideas—for example, the idea of distribution, the importance of specifying the kind of average, the concept of variability—are involved even in the description of a single sample. Chapter 5 stressed the importance, when interpreting any measurement of a given subject's behavior, of comparing it with similar measurements that have been made of some identifiable reference group.

Chapters 7 and 8 demonstrate that the description of a sample is frequently not an end in itself—that inferences can be made from what you know about a particular sample to what you would like to know about the population whence it came. Be aware, however, that description and inference are independent functions. It is true that inference, when it occurs, is always from sample to population; but description may, in principle, be applied directly to a population, and in many institutional settings, that is exactly what happens.

Those are all important ideas. They won't enable you to do social science research, but they will enable you to understand research done by others. That is clearly a high enough payoff to have justified the effort you have expended in mastering those ideas; but there is an additional reward that, though incidental to the stated objectives of the book, could well prove most important of all in the end. While learning to think better statistically, you may have learned to think better generally!

You may not have occasion to think statistically every day, every week, or even every month; but whenever the occasion does arise, you will be ready for it.

After an especially long hiatus, you may not be able to deal immediately with every concept that you have mastered while studying the book, but you will find that a quick reference to its discussion of a particular concept will revive your understanding not only of that concept, but of related ones as well. In fact, the habit of reviewing relevant sections of this book is a good one even while perusing another. Books on statistical methods, for example, are based primarily on concepts presented here.

General agreement is so far lacking on a single set of symbols to represent those concepts. For that reason, I have provided on pages 167–168 a list of commonly used statistical symbols that you may encounter in your reading.

Good luck! And remember what Sir Francis Galton used to say: "Whenever you can, count."

List of Symbols

Entries in the table on the following page are listed in two categories:

1. Every "Our symbol" (i.e., a symbol used in this book) is included in columns 1, 3, and 5 in the table unless, like ρ (rank-difference correlation) and AD (average deviation), it has been described in the text only as an introduction to a more useful but more complicated concept (such as the standard deviation or the "Pearson r").

2. The list of "Other symbols" that you might encounter elsewhere may include the most common alternatives, but it is not exhaustive. There is no standard usage either accepted by authors or endorsed by the American Statistical Association.

| Sample | | Population | | Sampling Distribution | | |
Our symbol	Other symbol(s)	Our symbol	Other symbol(s)	Our symbol	Other symbol(s)	Verbal designation
X	x	X	x			Raw score
n	N	N	n			Size (number of observations)
$\overline{X}$	$\bar{x}, M$	μ	$\overline{M}, \hat{M}$			Mean
Mdn						Median
x_{sample}	$X - \overline{X}, d$	$x_{pop.}$	$X - \overline{X}, \overline{d}, \hat{d},$			Deviation score
S	s, SD, σ	σ	$\overline{\sigma}, \hat{\sigma}$			Standard deviation
S^2	V					Variance
		s	$\overline{s}, \hat{s}$			Estimated standard deviation
		s^2, MS				Estimated variance (mean square)
				$s_{\overline{X}}$	$s_{\overline{X}}, \hat{s}_{\overline{X}}$	Estimated standard error of the mean
				$s_{\overline{X}_1 - \overline{X}_2}$	$\overline{s}_{\overline{X}_1 - \overline{X}_2}, \hat{s}_{\overline{X}_1 - \overline{X}_2}$	Estimated standard error of the difference between means
		z				z ratio
				t		t ratio
				F		F ratio
				χ^2		Chi-square
r		r				Pearson product-moment coefficient of correlation

Notes

Chapter 3: Measures of Central Tendency

1. If you should ever have to do the computations yourself, you would find that the median is often within a class interval rather than precisely between two intervals as in the illustrations I have cited. In that situation, it is necessary to interpolate; almost any text on statistical methods will quickly tell you how.

Chapter 4: Measures of Variability

1. Every score is considered to be at the midpoint of an interval 1 unit wide, so the interquartile range actually extends from $\frac{1}{2}$ unit below the score at Q_1 to $\frac{1}{2}$ unit above the one at Q_3. As with the median, interpolation is necessary if either quartile falls within a class interval larger than 1.

2. With respect to the additional $\frac{1}{2}$ unit on either end, the same is true of the total range as of the interquartile range (see note 1 above).

Chapter 5: Interpreting Individual Measures

1. Jum C. Nunnally, *Psychometric Theory* (New York: McGraw-Hill, 1967), p. 2. Measurement by this definition specifies one type of scale—its numbers are called *cardinal* because they are generally preferred, but most authorities recognize two other types: *nominal,* in which data are merely classified, and *ordinal,* in which the classes are placed in order of size, ability, prestige, or some other attribute that can be ordered. Some entities, like genders or countries of origin, can be classified but cannot be ordered. Others, like military ranks or beauty contest awards, can be not only classified but ordered as well. Still others, like grade point or batting average, can be not only ordered but also placed on a scale of constant units, like grams, centimeters, number of hits in a baseball season, and so forth. We prefer this last type—a quantitative scale—but we have ways of dealing with the other two. (See also note 1 for Chapter 10.)

2. Originally a *T* score was any standard score other than a *z* score, but usage has gradually changed its meaning to "a standard score in a distribution that has a mean of 50 and a standard deviation of 10."

3. In the former case (mental age = 6), the child's IQ is 100. This is obtained by dividing the mental age (MA) by chronological age (CA):

$$\frac{MA}{CA} = \frac{6}{6} = 1.00$$

and then multiplying by 100 to get rid of the decimal point. In the latter case, the child's IQ is 133:

$$\frac{MA}{CA} = \frac{8}{6} = 1.33$$

The IQ defined above was for many years *the* IQ. Now, however, it is known as the "ratio IQ" to distinguish it from the newer "deviation IQ" (see "Deviation IQs" scale in Figure 5-5 on page 57). A deviation IQ is a standard score. In a distribution of general mental ability test scores, the obtained mean is converted to 100, because the mean ratio IQ is 100. Then the standard deviation of the raw scores is changed to match that of ratio IQs on the same test. The result is a set of standard scores that are almost the same as scores obtained by the old method. However, the new method is easier to use and has other technical advantages the explanation of which would require a more extensive digression into psychometrics than would be appropriate here.

Chapter 6: Correlation

1. Actually, a Pearson *r* coefficient on these data would be somewhat less than 1.00. The relationship is perfect, but *r* is accurate only in rectilinear (straight-line) regressions, and this one is really curved. (For every height increase of a uniform amount, the corresponding weight increase is systematically greater as you move from left to right in Figure 6-1.) Another coefficient can be used in such cases; look for "eta (η) coefficient" or "correlation ratio" in any text on statistical methods.

2. If you *should* ever need a quick estimate of correlation, rho is the one to use. Consult any standard text on statistical methods about what to do with tied ranks, and within a few minutes you'll be ready to use Formula (6-1).

3. In practice, the errors caused by such a loss of information tend to cancel each other, and unless there are many tied ranks, rho and *r* are almost always nearly identical. When they are not, *r* is preferred. [See R. P. Runyan and A. Haber, *Fundamentals of Behavioral Statistics,* 7th ed. (New York: McGraw-Hill, 1991), pp. 201–206.]

4. Those assumptions are (1) that each of the two distributions is unimodal and symmetrical and (2) that the line best representing the relation between them is straight rather than curved. (See the diagrams in the subsection on scatterplots in this chapter for examples of such *rectilinear* relationships. Also see note 1, above, concerning curved ones.) Actually, computer simulations [L. L. Havlicek and L. Peterson, "Effect of the Violation of Assumptions upon Significance Levels of the Pearson *r*," *Psychological Bulletin* 84, no. 2 (1977): 373–377] have shown that violation of the traditional assumptions does not affect *r* very much. It is therefore said to be a *robust* statistic.

Chapter 7: Description to Inference: A Transition

1. This is the standard explanation. I believe that I have discovered a flaw in it, but I have not developed another to take its place. A mathematical rationale for the standard explanation is nicely stated by Helen M. Walker, "Degrees of Freedom," *Journal of Educational Psychology* 31(1940): 253–260.

Chapter 8: Precision of Inference

1. It is possible conceptually to separate errors of sampling from those that inhere in the process of taking the measurement, such as allowing distracting sounds to enter the examining room. Our concern here is strictly with sampling error.

2. The number of individuals in an actual population is frequently unknown, and the N of a hypothetical distribution of means is infinite; there is thus no way in which either can be represented accurately in a drawing. But because I believe that the relations among samples, populations, and hypothetical distributions are best understood when presented graphically, I have drawn them anyway. The drawing of the population is larger than that of a sample because the sample is a part of the whole population, but the *amount* of the difference between them can never be known because we can't know the size of the population; my rendition is arbitrary in that respect. In Figure 8-1 the width of the drawings of sample means is smaller than that of the population because means vary less than individual scores

do. But the *height* of the distribution of sample means is infinite, so I have arbitrarily rendered its height equal to that of the population.

3.   The *shape* of the population in Figure 8-1, page 98, is normal, as is frequently the case in real-world applications. But it is not *always* the case. You might expect that a different configuration in the parent population would result in a distribution of sample means that also deviates from normal. You would be wrong. The *central limit theorem* states that *if in an infinite number of samples all were large and randomly obtained, the distribution of their means would be normal regardless of the shape of the parent population.* Here's why.

Remember the toy soldiers that I used to illustrate correlation? The correlation was between height and weight; concentrate now on the *height* dimension, and assume that every soldier is the same shape as every other, as in Figure 6-1, page 62. Now let's look at the entire population of toy soldiers manufactured by the TRU company: 20,000 units of each of 5 sizes, 100,000 in all. The distribution of the parent population is therefore not normal; it is rectangular.

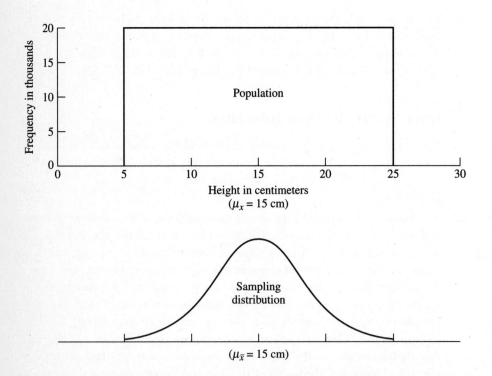

Since each score (height) is as likely as any other to turn up in any sampling of this population, the most likely shape of any sample is rectangular.

If a sample of $n = 5$ were to mimic the population exactly, there would be one soldier of each height, and their mean would be $\overline{X} = 15$—that is, precisely at the mean of the original 100,000 soldiers (the parent population).

But, of course, if we do actually calculate the means of many random samples of 5 soldiers, few of them *do* exactly mimic the parent population. When the samples are much larger than 5 we still see very few distributions shaped like the parent population; but many of their *means* will be precisely $\overline{X} = 15$ cm, few will be very far from that, and almost none will be found at the extreme scores of 5 or 25. In short, the distribution of *means*, given large enough samples, will be normal even when the population is not.

Chapter 9: Significance of a Difference between Two Means

1. This is the formula used when means are uncorrelated. If students were matched on some variable (for example, intelligence) related to their level of performance under both control and experimental conditions, a correlational factor would have to be introduced. But that refinement is beyond the scope of this book, and the basic idea of $s_{\overline{X}_e - \overline{X}_c}$ is better conveyed by Formula (9-1) as it stands.

2. To find how many standard errors an obtained difference is from zero, it is necessary to find the size of the standard error. In this instance, there were two such computations:

    First, there was the standard error of the difference for Figures 9-6 and 9-7, assuming that the standard errors of the means of the two distributions were as indicated in Figures 9-4 and 9-5:

$$s_{\overline{X}_c} = 2 \qquad s_{\overline{X}_e} = 2$$
$$s_{\overline{X}_e - \overline{X}_c} = \sqrt{s_{\overline{X}_c}^2 + s_{\overline{X}_e}^2}$$
$$= \sqrt{2^2 + 2^2} = \sqrt{4 + 4} = \sqrt{8}$$
$$= 2.83$$

(You will find this calculation also in Box 9-1.)

    Second, there was the standard error of the difference for Figures 9-10 and 9-11, assuming that the standard errors of the means of the two distributions were as indicated in Figures 9-8 and 9-9:

$$s_{\bar{X}_c} = 5 \qquad s_{\bar{X}_e} = 5$$

$$s_{\bar{X}_e - \bar{X}_c} = \sqrt{s_{\bar{X}_c}^2 + s_{\bar{X}_e}^2}$$

$$= \sqrt{5^2 + 5^2} = \sqrt{25 + 25} = \sqrt{50}$$

$$= 7.07$$

3. In our examples, the no-difference, or no-deviation, hypothesis specifies no deviation from a difference of zero between (or among) treatment groups. By extension, the null hypothesis becomes *any hypothesis concerning the location of a parameter*—a hypothesis of no difference *from that hypothetical location,* whether or not the location is zero.

    A good illustration of a nonzero null hypothesis comes from quality control in industry. If a product is supposed to contain X amount of some chemical, for example, then each random sample is drawn against a null hypothesis of X, and any significant deviation from X requires remedial action. (In this example only one sample is drawn, and the hypothetical distribution is of means rather than differences between means.)

4. The significance level states the probability that we are making an error when we reject the null hypothesis. That kind of error is known as *Type I.* The probability of making a Type I error is sometimes called "alpha error" or simply "alpha" (α). By stating before the experiment that we will accept only a very low probability (that is, a low probability of rejecting a null hypothesis that is in fact true), we can reduce our Type I errors to a level approaching zero. Unfortunately, however, the *lower* we set *that* level, the *higher* the probability of *accepting* a null hypothesis that is in fact *false.* The latter kind of error is called *Type II,* or "beta" (β). The following table summarizes those relationships. The ability of a significance test to make the decision represented by the lower left cell of the table—that is, to reject a false null hypothesis—is called its *power.* ("Sensitivity" might have been a more descriptive term for the power to detect a real difference.)

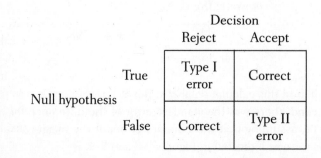

		Decision	
		Reject	Accept
Null hypothesis	True	Type I error	Correct
	False	Correct	Type II error

5. Many authorities will not accept a one-tail test under any circumstances. Their argument is that we cannot have a good reason for expecting one group to be superior to the other—or rather, that if we did, we wouldn't have to make the statistical test. To put it another way, the reason that we conduct the test at all is precisely that we *don't* know what the outcome will be. There is controversy in every field.

Chapter 10: More on the Testing of Hypotheses

1. Actually there is a procedure even more rigorous than specifying distances on a scale, but in the social and behavioral sciences it embodies a rigor that is almost never possible to attain directly. It is called a *ratio* scale. Measurement experts speak of four levels of measurement:

A. *Nominal.* All you do is *classify* things as described in the section about chi-square. Differences are in *kind* (qualitative), not *amount* (quantitative).

B. *Ordinal.* Classify, but then *place in order* as described in Chapter 6. Placement is quantitative.

C. *Interval.* Classify and place in order, but then specify the *distances* between points on a scale of *equal intervals;* e.g., the distance between a score of 7 and a score of 9 is the same as the distance between 76 and 78, or between 923 and 925.

D. *Ratio.* Classify, place in order, and specify distances between points of a scale, but then specify the distance of each point *from absolute zero.*

Social/behavioral scientists have developed many scales that approach the ideal of equal intervals; they have *not* often come anywhere near the ideal of a ratio scale, because their scales usually lack an absolute zero. (An IQ of 80 is not twice as high as one of 40, for example, because an IQ of zero represents not a complete lack of intelligence but only a failure to pass any of the items on the test.) Without an absolute zero it is impossible to discover whether one quantity is three times as large as another, or five times, or more generally to derive *any* ratio from data obtained on such a scale.

While very few of our data in the social/behavioral sciences come from ratio scales, we often *create* virtual absolute zeros. When we calculate the distance of each data point from the mean of a distribution, the "difference score" (x) of any data point that is exactly at the mean is zero, and it does represent a total absence of deviation from the mean. That is important, because it makes possible the multiplication, division, square, and square root operations on deviation scores that have been ubiquitous in this book since the end of Chapter 2.

2. Try to recall your recent encounter with the *t ratio*. If you need to refresh your memory, turn back to pages 117 ff. Pay special attention to Figure 9-12; there you will see three differently shaped distributions of *t* ratios. Now, a depiction of chi-square sampling distributions would look very different from that of *t* ratios, but the two have one important thing in common: Each has a differently shaped sampling distribution for every value of degrees of freedom. The tables for finding the *p* value of any given *t* ratio or chi-square reflect that fact, so before you can use those tables you must ascertain the *df* in your calculation. A good methods text will show you how to do that.

3. This latter estimate is sometimes referred to as *residual* or even *error variance* (s_E^2). The term *error variance* may be somewhat confusing, because most of the variance probably stems from factors that in different circumstances might prove to be effective independent variables. They are regarded as errors in the context of a given investigation because they are random with respect to the independent variables in *that* experiment. They are the so-called controlled variables. There are two general strategies for "controlling" a variable in an experiment: (1) to hold it constant over all conditions of the independent variable(s) and (2) to randomize its variance. Both strategies are designed to achieve the same objective: prevention of any systematic (i.e., *non*random) relationships between the controlled variable and any independent variable. There usually are several controlled variables in an experiment.

4. When only two groups are being compared, the analysis of variance is really a *t* test; to put it another way, *t* is a special case of analysis of variance that can be used only when there are just two categories of the variable being examined. In such a case, $t = \sqrt{F}$; with whole populations, of course, the same is true of *z*—that is, $z = \sqrt{F}$.

5. The *F* ratio for interaction is

$$F = \frac{s_I^2}{s_w^2}$$

where s_I^2 is the interaction variance and s_w^2 is the within-group variance.

6. In order to maximize simplicity—and hence clarity—in this example, I have assumed the linearity of each of the functions depicted in Figures 10-4 and 10-5. A more refined experimental design might have defined those functions more precisely by including several degrees of stress (instead of just two) and several degrees of experience. If that were done in the case of stress, for example, it might turn out that medium amounts of stress produce more persistence than either extreme amount. An investigation of the effects of a single variable through many changes in its value is sometimes referred to as a *parametric* study.

Chapter 11: Correlation, Causality, and Effect Size

1. There is now a purely statistical method of assigning causes in a matrix of correlations. It is called *cross-lagged panel correlation (CLPC)* and is based on the time relations within the matrix; it is axiomatic that causes precede effects [D. A. Kenney, "Cross-Lagged Panel Correlation: A Test for Spuriousness," *Psychological Bulletin* 82 (1975): 887–903]. There is no way that such an assignment can be extracted from a single correlation coefficient, however.

2. Even if a variable is really continuous, scores from it are reported in discrete intervals; even an interval of 1 breaks the continuum into segments. (For example, a score of 6 on a test is awarded to all persons whose performances are better than 5.5, but not as good as 6.5.) But *a dimension with a large number of class intervals approximates continuity,* so I shall refer to such data as *continuous.*

3. In the case of a qualitative difference (like gender), one would expect a normal distribution *within* each group rather than the truncated configurations (half of a normal distribution in each category) implied by Tables 11-1B, C, and D.

4. Of course the *r* must be reliable. Its reliability can be assessed by conducting a significance test. The test is similar to that of a difference between means (Chapter 9) except that the null hypothesis is different. Instead of "there is zero difference on the dependent variable between the two groups defined in the independent variable," the null hypothesis here is that "there is zero correlation between the independent and the dependent variables." In addition, instead of dividing the obtained difference between the two groups by the standard error of a difference between means ($s_{\bar{X}_e} - s_{\bar{X}_c}$), you divide the obtained *r* by its standard error $s_{r_{i.d.}}$, where *i* is the independent variable and *d* is the dependent. This works for coefficients of $r = .50$ and smaller. For larger *r*s, the testing of the null hypothesis is more complicated because the sampling distribution is distorted by the upper limit of 1.00 that applies to any coefficient of correlation.

5. A sophisticated technique is available for investigating causal relations among *naturally occurring variables* (as distinguished from the variables in an experiment). It is called *structural modeling, structural equation modeling,* or *path analysis.* Structural analysis differs from exploratory correlational research in that a structure (theory, hypothesis) must be clearly stated in advance. Like an experiment, such an analysis enables confirming data to render alternative structures implausible. A lucid overview of this kind of analysis is available in William E. Crano and Marilyn B. Brewer, *Principles and Methods of Social Research* (Needham, Mass.: Allyn & Bacon, 1986).

Solutions to Sample Applications

Chapter 3: Measures of Central Tendency

EDUCATION

A good choice. The mean of 200 reading scores will give you a reliable average of that ability in your trainees. It would also be useful later on to meet the need to estimate reading level in future 12th-graders, should the program be made permanent, thus turning your population into a sample of all district 12th-graders, current and future.

Possible misinterpretation. Misinterpretations include finding a single central tendency when there are really two. It is possible that students enrolled in the semiskilled trades (such as food service, construction, sewing, and horticulture) and students in the skilled trades (such as computer programming, electronics, and radio and television) have very different reading abilities. If so, the mean reading level of the total sample would fall between the two subgroups. It therefore would not represent either of them, and materials purchased on the basis of the mean would be at too high a level for one group of students and too low a level for the other. If the two groups were of approximately equal size, the resulting bimodal distribution could alert the investigator to that problem, but if one of the groups was much smaller than the other, the total distribution might be unimodal and thus give no clue to the presence of the second group. If that presence resulted in a skewed distribution and that distribution was treated as a single population, the appropriate measure of control tendency would be the median, which gives less weight to atypical scores.

POLITICAL SCIENCE

A good choice. The mean can give you an estimate of the balance point of the distribution of European military spending. Of all the measures of central tendency in your sample, the mean is the best estimate of the population average.

Possible misinterpretation. Despite the general virtues of the mean as a measure of central tendency, it is sensitive to any single deviant case. For example, if a 1980 sample had included the Soviet Union, the expenditure of that one country would have far exceeded that of any of the others and would therefore have skewed the distribution. The median would have been a more appropriate centrality measure, since the median divides a distribution in half. The Soviet Union would have had no greater effect on the location of the median than did any other nation in the upper half of the distribution.

PSYCHOLOGY

A good choice. The mean is the only measure of central tendency that includes information from all the children you have tested. It will yield the most reliable estimate of what the typical newborn is capable of doing (at least in your particular sample). If, for example, the mean number of points assigned were 50, then 50 would be a standard, or norm, for judging the normality of a particular infant, and a child scoring far below 50 could be viewed as abnormally delayed.

Possible misinterpretation. The neonates selected from the hospitals in your city may be different from neonates in general. For example, if your sample were composed mainly of African-American infants, the mean might be too high to be representative of infants in general because African-American infants tend to develop more rapidly than infants of other races.

SOCIAL WORK

A good choice. The mean can give you a reliable estimate of the average number of hours of service that families receive from this agency.

Possible misinterpretation. A single family that required a particularly intensive treatment program with many hours of service would raise the mean substantially unless n was very large. Even with a fairly large n, a few really extreme cases can distort the mean as a measure of central tendency. In general, the mean should be used only when the distribution is nearly normal.

SOCIOLOGY

A good choice. The mean can give you a reliable estimate of the average income of residents in the city.

Possible misinterpretation. The mean is sensitive to extreme scores. If the distribution included such scores and if they did not happen to be arranged so that they balanced each other, the mean would give you misleading information about the bulk of the population.

Chapter 4: Measures of Variability

EDUCATION

A good choice. The standard deviation can tell you whether the variability of the ratings for the two programs is the same. One program may be fairly uniformly rated by the great majority of teachers; that is, teachers may not differ very greatly in their assessments of the usefulness and practicality of the techniques presented in the program. The other training program may be rated very high on practicality and usefulness by some teachers and very low by others, resulting in a high standard deviation. The training program with the relatively large standard deviation may be less desirable because teachers may have widely varying perceptions about the usefulness and practicality of the techniques, leading to a less uniform adoption of the techniques than if the first program were used.

Possible misinterpretation.   Misinterpretations include attributing the higher variability of the second in-service training program entirely to problems inherent in the program itself. It may be due partly or even entirely to other factors—differences in the time of day the teachers were trained, for example, or differences in administrative support among schools.

POLITICAL SCIENCE

A good choice.   The standard deviation can tell you how much the incidence of military coups differs among Latin American countries. If each country experienced approximately the same number of coups, there would be almost no differences and the average of their deviations from the mean would be nearly zero. But the more differences there are, and the larger they are, the larger that average would be. The standard deviation is a kind of average of the deviations of individual countries from their mean.

Possible misinterpretation.   As in the case of the mean, a few extreme scores within a frequency distribution can give misleading results when the standard deviation is calculated. Therefore, if a very few Latin American countries experienced a very high number of coups, then just as the median would be a better measure of central tendency than the mean, so the interquartile range would be a better measure of variation than the standard deviation.

PSYCHOLOGY

A good choice.   The standard deviation provides a measure of consistency (agreement)—or rather, a measure of inconsistency (disagreement)—among the observers. If, for instance, some observers reported as few as 2 aggressive acts per day and others reported as many as 20, the standard deviation would be large, and the reliability of the ratings should be questioned.

Possible misinterpretation.   Given a high level of variability among the five observers in our example, it would be tempting to blame one or several raters (e.g., the least experienced) for the apparent discrepancies and inaccuracies. A more likely explanation would be that the instructions to the observers failed to clarify what constitutes an aggressive act. If so, the instructions would need to be revised.

SOCIAL WORK

A good choice.   The standard deviation can tell you the amount of variability in grant awards to member agencies. A high standard deviation would indicate considerable differences in the amounts of money received by the agencies. A low standard deviation would indicate that agencies were funded at approximately the same level.

Possible misinterpretation.   Again, as with the mean, a few extreme values can result in a higher standard deviation; thus, although only one or two agencies might be receiving much more or much less money than all the others, a large standard deviation would make it appear that there was much variation throughout the distribution. In general, the standard deviation should be used only when the distribution is approximately normal.

SOCIOLOGY

A good choice.   The standard deviation can give an estimate of that span of family sizes which includes 68 percent of all families and that which includes 96 percent of all families. Once you know the standard deviation, the regions above and below the mean can be examined separately by referring to normal distribution tables.

Possible misinterpretation.   All these conclusions are based on the assumption of a normal distribution of family sizes. If that distribution was not approximately normal, you would have to find other ways of reporting your conclusions. (See the section on "The Interquartile Range.")

Chapter 5: Interpreting Individual Measures

EDUCATION

A good choice.   If you transform raw scores for all the tests of the battery into standard z scores based on performances by students of the same age and grade, it will be easier for you to compare the student's performance on the various tests. For example, a z score of 0 would indicate that the student is average in that ability in comparison with his peers. A z score of $+1$ would indicate that the student is 1 standard deviation above the mean (at the 84th percentile) in that ability in comparison with fourth-grade students nationwide. A z score of -2 would indicate that he is 2 standard deviations below the mean (at the 2nd percentile) of his grade. Such comparisons can help the teacher to determine in which areas the student is having difficulty so that some specific assignments can be designed to strengthen those abilities.

Possible misinterpretation.   Misinterpretations include regarding ability scores as direct indices of hereditary potential. For instance, if a student is a member of a lower-class ethnic minority, his relatively low abilities may be due to environmental rather than hereditary limitations.

POLITICAL SCIENCE

A good choice.   The z score can tell you where each score fits on a scale common to all. The z scale is based on each distribution's mean and standard deviation. In this case, each of the various civil strife scores is expressed in terms of the standard deviation of its own distribution and measured from its own mean. Once all the scores have been converted to z scores, every distribution has the same mean (0) and standard deviation (1). *Now* it makes sense to combine them.

Possible misinterpretation.   Don't regard standard scores as absolute in the sense that physical measures (e.g., length) are absolute. Each score tells only where a given measure lies within one distribution of measures of the same attribute. In a different distribution, the same measure might yield a different standard score.

PSYCHOLOGY

A good choice.   By transforming each raw score to a new scale with a mean of 0 and a standard deviation of 1, you can ascertain whether June is developing normally in all three domains. If she earns z scores of −1 on all three tests, you can infer generally delayed development. If all three z scores are +1, you can infer advanced development. If she obtains z scores of, say, +2 on intelligence, −1 on social development, and −2 on psychomotor development, you can infer fragmented development.

Possible misinterpretation.   One possible misinterpretation is the presumption that obtained measurements are stable across time. Actually, the behavior of infants and young children tends to vary considerably from one observation to the next.

SOCIAL WORK

A good choice.   Someone in the personnel office can place each raw score into a distribution with a mean of 0 and a standard deviation of 1. Then all scores are on a standard scale, thus making it possible for you to compare them.

Possible misinterpretation.   If the various tests have been standardized on *different* populations, comparisons will be risky. For example, scoring high on client assessment in a population of beginners may not be more laudable than scoring low on client treatment in a population of highly trained professionals.

SOCIOLOGY

A good choice.   The z score can tell you how this professor compares with other professors in terms of standard deviation units above or below the mean.

Possible misinterpretation.   Professors teaching the courses for which you are eligible may differ in some significant way from the faculty as a whole (e.g., they may have been selected for their ability to relate to lower-division students). If so, the professor's z score may well lead you to make a bad decision at registration time. For example, his z score on authoritarianism could be low in the general faculty but high among those faculty members who are teaching the course in question.

Chapter 6: Correlation

EDUCATION

A good choice.   The Pearson r can tell you whether there is a relationship between self-concept and social responsibility in these children. It will also tell you whether the relationship between self-concept and social responsibility is positive or negative. The size of the coefficient, whether positive or negative, will indicate the strength of the relationship between the two measures.

Possible misinterpretation.   From a strong correlation (say, .7 or .8), it might be erroneously concluded that a student's self-concept causes the student to be socially responsible or irresponsible. (Although the two variables do vary together, this *may* be due to some third variable, such as previous school or home experiences, which affects both social responsibility and self-concept.) It might also be erroneously concluded that a program effective in enhancing a person's self-concept necessarily increases social responsibility. (If self-concept is caused by social responsibility or if both are caused by some third factor or set of factors, changing self-concept will not affect social responsibility.)

POLITICAL SCIENCE

A good choice.   The Pearson r can tell you the magnitude and direction of the relation between the amount of conflict within countries and the amount of foreign conflict they initiate. First, the larger the coefficient, the stronger the association. Second, a positive sign indicates a direct relationship between internal and external conflict, while a negative sign indicates an inverse relationship.

Possible misinterpretation.   Normally, a strong positive value of r can be taken as evidence that domestic conflict is associated with foreign policy conflict. Remember, however, that *correlations* do not necessarily entail *causal* relations. Also be aware of the assumptions behind r. If these assumptions are violated, then the results may not be valid. The most important of these assumptions are that the relationship between the two variables yields a straight regression line, that the distributions are unimodal, and that they are fairly symmetrical. However, violation of even these assumptions seldom invalidates an r. (See note 4, this chapter.)

PSYCHOLOGY

A good choice. The Pearson r will indicate the degree of association (.00 to either +1.00 or −1.00) between sugar intake and activity rating for the 100 children. The relationship can be positive (high sugar levels associated with excessive activity and low sugar levels associated with low activity) or negative (high sugar levels associated with low activity and low sugar levels associated with high activity).

Possible misinterpretation. A significant positive or negative relationship does not imply a cause-and-effect relationship. A positive Pearson r does not *necessarily* mean that high sugar levels cause hyperactivity. Hyperactivity may stem from a fundamental impulsivity in children. That is, the child's inability to control impulses may lead to both excessive activity *and* excessive ingestion of sugar. Thus a third factor (impulsivity) may be responsible for an observed relationship between blood sugar and hyperactivity.

SOCIAL WORK

A good choice. The Pearson r provides a measure of association that reveals both the strength and the direction of the relationship between the two variables. Your hypothesis is that women of high self-esteem are less dependent on welfare than are women of low self-esteem. A high negative correlation would confirm your hypothesis.

Possible misinterpretation. Even if your hypothesis is confirmed and you know self-esteem and requesting welfare are related, you don't know just *how* they are related. Two possibilities are (1) that self-esteem renders a woman more determined to make herself economically self-sufficient and (2) that an inability to make herself self-sufficient damages a woman's self-esteem.

SOCIOLOGY

A good choice. The Pearson r can tell you whether there is a relationship between the conservativism of academic disciplines and the authoritarianism of their professors.

Possible misinterpretation. Misinterpretations include the notion that the first variable causes the second, or vice versa. In fact, both variables could be products of an unidentified external factor or set of factors, or both could be parts of a larger system that requires them to vary together.

Chapter 8: Precision of Inference

EDUCATION

A good choice. The standard error of the mean is used to establish a confidence interval within which the population mean for each grade level lies. The band of scores described by the confidence interval provides a better index than the obtained mean itself of the achievement level of each local grade because it takes into account the errors that inevitably occur whenever measurements are made. You can give the probability that the mean really lies between two particular values.

Possible misinterpretation. If the band provided by the confidence interval around a sample mean is higher or lower than the national mean, don't assume that the difference is due solely to school programs. Other factors such as the influences of the home and the community must also be considered.

POLITICAL SCIENCE

A good choice. Precisely these limits are described by the confidence interval. There is a confidence interval for any desired probability (confidence level). For example, say you want to specify the interval within which you can place the true mean at a confidence level of 68 percent (i.e., the interval within which the probability is 68 percent that the true mean lies). Your obtained mean is 50. If the standard error ($s_{\bar{x}}$) is 10, the interval is $2 \times 10 = 20$ points wide (see pages 100 ff). At the 68 percent level of confidence, then, the confidence interval would extend from 40 to 60 points.

Possible misinterpretations. You might think it equally probable for the true mean to be at any point within the confidence interval. Actually, the probability is higher in the middle of the interval than anywhere else. Your obtained mean is, after all, your best single-point estimate of the true mean.

PSYCHOLOGY

A good choice. The confidence interval would allow you to estimate the limits within which the boy's true inkblot test score resides, and the confidence level would provide a statement of the probability that his true score is indeed within those limits. If the limits were far apart or the probability low, you would be cautious in interpreting the boy's score.

Incidentally, the manual might include, either in place of the confidence interval or in addition to it, a correlation coefficient (r_{xx}). That statistic represents another way of looking at reliability. (See Chapter 6.)

Possible misinterpretation. You may be tempted to think of error as something that lowers a score. But measurement error (unreliability) can also raise a score. The confidence interval therefore extends both below and above the obtained score. (The obtained score is in the middle of the interval.)

SOCIAL WORK

A good choice. The confidence interval provides a range of scores within which the family's true FLIP rating probably resides. There are two ways to approach the question "Where is the true score?"

First, you can set limits above and below the obtained score and then find the probability that the true score is between those limits. Second, you can set an acceptable probability and then set the limits that match it. In either case, the interval between the limits is called the confidence interval, and the probability is the level of confidence in this particular case.

Possible misinterpretation.   In tests of significance (see Chapter 9) the probability cited is that of an occurrence *outside* prescribed limits. Here, you report the probability that the true mean is *inside* those limits.

SOCIOLOGY

A good choice.   The standard error of the mean can give you a band of scores (the confidence interval) that probably contains the true mean. It can also specify the level of that probability (the level of confidence). This information is far from "meaningless."

Possible misinterpretation.   The true mean does not *have* to fall inside the intervals you have computed—thus, you shouldn't accept a confidence interval without mentioning the corresponding level of confidence.

Chapter 9: Significance of a Difference between Two Means

EDUCATION

A good choice.   The *t* ratio can tell you whether at the end of the semester the sample of students who were in the group-counseling program differs significantly in teacher-rated disruptiveness from the sample that was not in the program. Thus the staff could determine (within a given probability) whether students in the group-counseling program were less disruptive at the end of the semester than the students not in the program.

Possible misinterpretation.   You can't necessarily attribute differences between the groups solely to the counseling program. Other factors may have influenced the results:

1. Even though the groups were randomly assigned, there may have been initial differences in disruptiveness at the beginning of the semester. The probability of that occurring is inversely proportional to the *sizes* (*n*s) of the groups. Groups of 50 probably would not pose much of a problem in this regard.

2. The teachers knew which students were in the treatment program, and that may have influenced them to react to those students differently or to rate them differently even though those students actually were not different from those who were not in the program.

3. Just the added attention of the school counselor or just their inclusion in a new program could have lowered the students' disruptiveness, regardless of the content of the program.

POLITICAL SCIENCE

A good choice. The *t* test can be used to determine whether the observed difference between the means of the two groups is due to chance. If it is, the average difference between the two means within repeated pairs of random samples taken from the same population will be zero. If the observed difference is large enough to be statistically significant, then we can reject the hypothesis that the difference is merely a chance difference between two random samples of the same population of precincts. We can conclude rather that the precincts that participate in the new program probably are truly different from those in the old cops-only program.

Possible misinterpretation. Although the *t* test is appropriate for analyzing the difference between the means of two small samples from the same or identical populations, its use is subject to three restrictions. First, the observations in the two samples must be independent of each other. Second, the populations must not be skewed in opposite directions. Finally, if the populations do not have equal variances, adjustments are needed in the calculation of *t*.

PSYCHOLOGY

A good choice. The *t* ratio can tell you whether the relaxation treatment or the drug therapy is the more effective as determined by the mean scores of the two groups. It will also give you the probability that there really is no difference between them—that they are both random samples from a single population with respect to activity level.

Possible misinterpretation. Using a one-tail test here would be a mistake because you have no reason to claim superiority for either treatment. Even if a two-tail test reveals a significant difference between the two groups, locating the cause of that difference may be difficult. Suppose that the drug-therapy group appears to benefit more from treatment. That could be the result of the drug, but unless you equalized the age of the two groups, age differences could be the critical factor. Without controlling for other possible causes, you cannot know whether the cause is the drug or some other factor such as the ages of the children, their intelligence, their social background, or some characteristic of their drug therapists.

SOCIAL WORK

A good choice. The *t* ratio can be used to determine whether there is a significant difference between two groups. The difference is significant if you can reject the null hypothesis that the two groups are from the same population with respect to health. You can take your *t* ratio to a table that will give you the probability that the two groups *are* from the same population. If that probability is extremely small—say, .01—you may assert with considerable confidence that the experimental program has been effective.

Possible misinterpretation. Your confidence in the program will have been misplaced if some outside factor influences one group but not the other. For example, if the experimental group rides a special bus to the center, eats its meals together, or does something else together as a by-product of the program, it could be all that "togetherness" rather than the program itself that improves the seniors' health.

SOCIOLOGY

A good choice. The *t* ratio, entered into appropriate tables, can tell you how likely it is that the obtained difference has occurred by chance.

Possible misinterpretation. As in correlation, a relationship does not by itself justify an inference of causality. If most of the Catholics in your state belong to a different social class than most non-Catholics, it might not be legitimate to infer that the difference in religious belief is the cause of the difference in family size. It might turn out that with social class held constant, Catholic families are no larger than non-Catholic ones.

Chapter 10: More on the Testing of Hypotheses

EDUCATION

1. *A good choice.* You need to know whether the frequency of high school graduation is higher for students who have been through the special program than for those who have not. Chi-square can tell you that.

Possible misinterpretation. It is important that all the students in both groups come from the at-risk category. A group taken from the entire student body would have an initial advantage over a group identified as being at risk. Also, you cannot be certain that the training program is the only cause of the observed difference. Even if the program significantly increases the percentage of at-risk students who graduate, its success may be due more to the interest of the staff in a new program than to any attribute of the program per se.

2. *A good choice.* One-way analysis of variance can tell you whether the differences among the three means on the social-problem-solving test is greater than would be expected by chance. If there is such a difference (and it is in favor of the clients of the trained counselors), then you may conclude that the students who went through the program are better able to solve the problems posed by the test than are students in the other two groups.

Possible misinterpretation. Don't assume that if there is a significant F ratio, each group mean differs significantly from each of the others. For the F test to be significant, it is necessary only that two of the means differ significantly. You may also erroneously conclude that the test result must be attributable to the program. Even though the students were randomly assigned, there could have been differences among the groups before the programs began (though that is unlikely if the groups are large). The students could have learned which groups they were in after the study began and been affected by that knowledge. Similarly, each counselor's knowledge of his or her own place in the program might affect student behavior in ways not specified by the training the counselor received.

3. *A good choice.* Two-way analysis of variance of the abstract-reasoning-ability scores from the beginning of the year can tell you whether there is any difference among the mean test scores of the students in the four groups. The same analysis can tell you whether there is a significant difference in reasoning ability between the sixth and seventh grades prior to your intervention. The analysis can also tell you whether, before the programs begin, there are significant interactions between grade level and abstract reasoning ability in the four treatment programs.

At the end of the year, all those significance tests can be administered again. In addition, and probably more important, if the differences among the means of the four groups were not significant before the training, a significant F test of program differences afterward suggests differential effects of the four treatment conditions.

Possible misinterpretation. Attributing significant differences at the end of the year to the effectiveness of one or more programs would be an error if the differences were actually there in the beginning.

POLITICAL SCIENCE

1. *A good choice.* Chi-square can tell you the probability that any deviation of the observed frequencies from a stipulated expected frequency is due to chance. (The null hypothesis here is that equal proportions of Republicans and Democrats support a cut in the capital gains tax.) Chi-square compares the observed frequencies in each cell of the contingency table with what would be expected in these cells if the two variables—party affiliation and support for the capital gains tax cut—were independent.

Possible misinterpretation. The most common misuse of chi-square is the violation of certain key assumptions. The most important of these is that the data represent a random sample of independent observations. In this case, you must be sure that you sample *all* Republicans, not just some subgroup that is especially dedicated to the passage of a tax cut, and that your sample of Democrats is similarly random.

2. *A good choice.*   One-way analysis of variance can tell you whether observed differences (in number of coups) among types of regime are likely to have occurred by chance. The analysis of variance is an extension of the difference-of-means *t* test that we examined earlier. Had there been only two types of regime legitimacy, a *t* test would have sufficed. With more than two types, however, you must compute an *F* ratio. If the population variance in coup frequency estimated from differences between the countries is not significantly greater than the variance as estimated from within the groups, then you cannot reject the null hypothesis. That is, you must conclude that the differences you have observed are merely random variations — that the three types of country are all one with respect to number of coups. (Note that "number of coups" is here treated as a *score* in the same way that "number of correct answers" on a test is treated as a score. "Number of persons favoring capital gains tax," on the other hand, as illustrated in the question immediately preceeding this one, is treated not as a score but as a *frequency.*)

Possible misinterpretation.   When the requirements of the one-way analysis of variance are met, the *F* ratio allows you to determine whether the observed difference among the group means is large enough to be statistically significant. However, since in this case there are more than two categories within an independent variable, the *F* ratio does not tell you which of the categories differ. Further tests are necessary to accomplish that.

3. *A good choice.*   Two-way analysis of variance can tell you whether there is a significant "form of government" main effect and whether there is a significant "type of leadership" main effect. It can also tell you whether there are any interaction effects. For example, it might turn out that totalitarian governments are more aggressive than other forms only when their leadership is of the unitary type.

Possible misinterpretation.   Two-way analysis of variance is subject to the same kinds of limitations as the one-way analysis of variance. Specifically, the *F* test does not tell you precisely where among the nine categories the significant differences lie. Again, further tests can be made.

PSYCHOLOGY

1. *A good choice.*   Chi-square can tell you whether the number of children who pass the test after treatment is greater or smaller than the number that might be expected to pass without treatment. If previous research indicates that 45 percent of untreated agoraphobics recover spontaneously, this figure might be used as the expected value (the null hypothesis).

Possible misinterpretation.   In this design, uncontrolled factors could account for significant findings. For example, just coming to the clinic may be sufficient to induce change, or children who come to the clinic may not be a random sample of agoraphobic children generally.

2. *A good choice.*   One-way analysis of variance can tell you whether there are significant differences among the three treatments in the children's self-ratings of experienced pain.

Possible misinterpretation.   You might infer from the F ratio that every mean differs from every other, but a significant F ratio tells you only that there is a difference somewhere between or among treatment means. You now have to scrutinize your data to ascertain where any differences are.

3. *A good choice.*   A two-way analysis of variance allows you to test: (1) whether any of the four group means (introvert–individual, introvert–group, extrovert–individual, and extrovert–group) is significantly different from any other; (2) whether group therapy or individual therapy is the more effective, regardless of introversion–extroversion tendencies; (3) whether introversion–extroversion tendencies are associated with better outcome, regardless of the type of therapy; and (4) whether there is an interaction between introversion–extroversion and therapeutic modality—that is, whether individual therapy works best with introverts and group therapy works best with extroverts, as you had predicted.

Possible misinterpretation.   In this particular example, there are only two categories of each variable (introvert-to-extrovert and individual-to-group). Whenever that is true, the F ratio for each variable identifies precisely the source of any difference that we find. For example, if the F for the introvert-to-extrovert variable turned out to be significant, we would know that the difference is between the group we have designated "introvert" and the one that we call "extrovert."

When there are more than two categories of a variable, however, the F test does not identify the source precisely. If, for example, we had divided our introvert-to-extrovert variable into three categories ("introvert," "ambivert," "extrovert") instead of two, a significant F would mean only that there was a difference somewhere within that variable. It would not tell us whether that difference was (1) between introverts and ambiverts, (2) between introverts and extroverts, or (3) between ambiverts and extroverts. Further tests would be necessary for more precise identification.

SOCIAL WORK

1. *A good choice.*   Chi-square can tell you whether the frequency of recidivism is significantly lower in the group of youth appearing before the board than in the group who are sent to court. Chi-square in this application is an index of deviation from the frequencies you would expect if the new board's procedures were not any more (or less) effective than the courts'. Indeed, chi-square *is* that deviation, squared and expressed as a proportion of the expected frequency.

Possible misinterpretation.   Chi-square deals with frequency counts only. Every score is therefore required to be either 0 or 1, and information about the magnitude of offenses is lost. It is *possible* that the relatively few offenders who break the law in spite of the board's rehabilitation efforts are guilty of offenses more serious than those committed in larger numbers by the control group.

2. *A good choice.*   One-way analysis of variance can tell you whether there is a significant difference anywhere among the four groups.

Possible misinterpretation.   A significant F does not mean that every group is different from every other with respect to hyperactivity. It tells you only that there is a difference somewhere among the groups. If the F test does reveal a significant difference, you must then follow up with other tests especially designed for use after the F test. Those tests will locate the differences.

3. *A good choice.*   Two-way analysis of variance can tell you whether there are significant differences among the four group means, that is, (1) children–foster family, (2) youths–foster family, (3) children–group home, and (4) youths–group home. Two-way analysis of variance also identifies the *location* of any effects—that is, whether adjustment is affected by age, type of foster care, or their interaction.

Possible misinterpretation.   A 2×2 factorial design yields an F ratio for each main effect and the interaction, so it is *not* subject to the ambiguity described earlier for one-way analysis of variance with six categories of the independent variable. The location of any difference is known from the first computation, because with only two categories on either variable the difference detected by the F test can only be between those two.

Sociology

1. *A good choice.*   Chi-square can tell you whether the numbers in the cells of your 2×2 table (early versus late $\times$ approval versus disapproval) could have occurred on the basis of chance alone. The other possibility is that these numbers were influenced by something other than chance. At least one such influence, if chi-square is significant, is belief about when human life begins.

Possible misinterpretation.   Possible misinterpretations include regarding the chi-square test as a direct answer to the general question implied by your hypothesis. The general question is "Does a person's belief about when human life begins affect his attitude toward abortion?" You might get a different answer if, for example, "early" and "late" were defined in relation to "nine months" instead of "90 days." Or your respondents' attitudes toward abortion might look different if question 2 were phrased ". . . under any circumstances" instead of ". . . on demand." Such wording might also affect the answer to the general question.

2. *A good choice.*   If "family size" is treated as a *score*—a quantitative attribute of families—one-way analysis of variance will give you the probability that all

the obtained differences occurred by chance. If that probability is very small, at least one of the differences is statistically significant.

Possible misinterpretation.   A significant F does not mean that each sample is significantly different from every other. The F test indicates only that there is at least one significant difference among those identified. If the F test is positive, then each mean must be compared with every other mean, and each difference must be evaluated separately from the others. But you may not use a t test in these circumstances. (See "Afer the F Test," pages 138–139.)

3. *A good choice.*   Two-way analysis of variance is an appropriate method. The summary table of F ratios will tell you whether there is a main effect of church attendance, a main effect of educational level, and/or an interaction between the two.

Possible misinterpretation.   In this particular example, there are only two categories of each variable (church attendance–nonattendance and high education–low education). Whenever that is true, the F ratio for each variable identifies precisely the source of any difference that we find. For example, if the F for the attendance–nonattendance variable turned out to be significant, we would know that there probably is a difference between the group we have designated "church attendance" and the one we call "nonattendance." The same is true of the education variable and for church–education interaction.

When there are more than two categories of a variable, however, the F test does *not* identify the source precisely. If, for example, we had divided our attendance–nonattendance variable into three categories ("high," "medium," and "low" attendance) instead of two, a significant F would mean only that there was a difference somewhere within that variable. It would not tell us whether that difference was (1) between "high" and "medium," (2) between "high" and "low," or (3) between "medium" and "low" attendance. Further tests would be necessary for more precise identification.

Chapter 11: Correlation, Causality, and Effect Size

Some possible alternatives stated in general terms: When Variable X and Variable Y are substantially correlated, maybe

X is causing Y; or perhaps

Y causes X; or possibly

both are caused by a third variable or set of variables; or even

each is but a single aspect of a multifaceted but indivisible whole.

An illustrative application of those alternatives.   For example, in the application to education (page 87) it may be that worthy self-concept is causing social responsibility; or perhaps

social responsibility causes worthy self-concept; or possibly

both are caused by a third variable or set of variables—maybe by midlevel socioeconomic status, with all the life experiences that attend that status; or even

worthy self-concept and social responsibility are but two of many aspects of a particular kind of personality.

A caution born of experience.   Take your choice among these alternatives, but do not invest much confidence in any of them until it has been tested by some means other than correlation. (See the discussion of correlational versus experimental studies on pages 151–153.)

Suggested Readings

Many valuable texts discuss how to *do* statistics and how to apply statistical principles to problems in the social/behavioral sciences and related professional endeavors. (If you are going to take a *course* in statistical methods, your instructor will make a selection for you.) Here is a very small sample of what is available:

GROEBNER, D. F., and SHANNON, P. W. 1994. *Essentials of Business Statistics: A Decision-Making Approach*, 2d ed. New York: Macmillan.

McCALL, R. B. 1997. *Fundamental Statistics for Behavioral Sciences*, 7th ed. Fort Worth, TX: Harcourt Brace College Publishers.

MOORE, D. S., and McCABE, G. P. 1999. *Introduction to the Practice of Statistics*, 3d ed. New York: W. H. Freeman and Company.

RUNYON, R. P., and HABER, A. 1995. *Fundamentals of Behavioral Statistics*, 8th ed. New York: McGraw-Hill.

WELKOWITZ, J., and EWEN, R. B. 1990. *Introductory Statistics for the Behavioral Sciences*, 4th ed. Fort Worth, TX: Harcourt Brace College Publishers.

WITTE, R. S. 1993. *Statistics*, 4th ed. Fort Worth, TX: Holt, Rinehart & Winston.

Index

In this index, where "ff" follows a page number, references to the listed topic can be found not only on that page but on subsequent pages within the same section. Each word or phrase signifies a topic but is not necessarily present in every place where that topic is discussed, and the listed pages are not necessarily the only ones on which listed items can be found. Summaries, calculation boxes, and Sample Applications are not included.